THE MEN WHO INVENTED BRITAIN

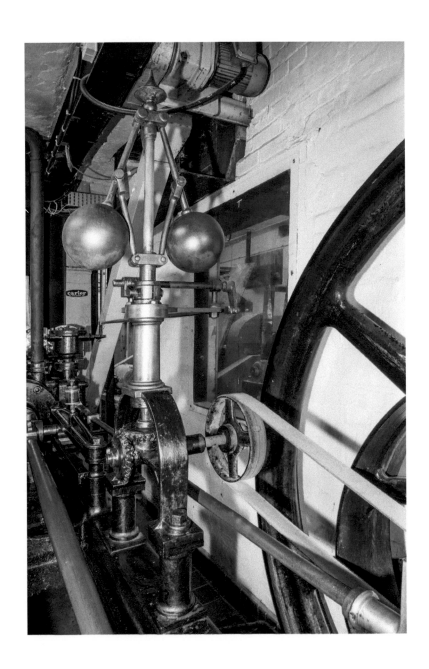

THE MEN WHO INVENTED BRITAIN

AN ILLUSTRATED INTRODUCTION
TO GREAT BRITISH ENGINEERS

JOHN HANNAVY

Whittles Publishing

Published by

Whittles Publishing Ltd,
Dunbeath,
Caithness, KW6 6EG,
Scotland, UK

www.whittlespublishing.com

ISBN 978-184995-568-3

Printed and bound by
CPI Group (UK) Ltd, Croydon, CR0 4YY

CONTENTS

ACKNOWLEDGEMENTS

The photographs in this book were all either taken by the author, or are from the author's collection of historic images, except those individually acknowledged.

In addition to Samuel Smiles whose writings inspired this project, I am, as always, indebted to the many experts and volunteers at today's industrial sites, museums and libraries, who have welcomed me, given me access to their collections, and shared their immense knowledge with me. Without them – and their willingness to dig out obscure historical references on my behalf, or let me explore their sites freely with my camera – this book would not have been possible. Amongst them are Chris Allen; Simon Barber at Claremont Controls Ltd.; Bal Bhelay, Senior Project Site Manager, Redrow Homes South East; Geoff Burns; Steven Campion at the British Library; Peter Dunn and the team at Claverton Pumping Station, Somerset; Peter Doré, Rachel Clare and colleagues at the Library of Birmingham; Jon Giles at Stone Mill Trowbridge; Kate Harland and Juan Cunliffe at Lion Salt Works, Cheshire; David Johnson at the Smeatonian Society of Civil Engineers; John Phillp and colleagues at the Northern Mill Engine Society, Bolton; Charlotte Robson; Colin Saxton at the Moseley Heritage Museum, Redruth; Angela Schad at the Hagley Museum and Library in Wilmington, Delaware; Leofric Studd at the Garlogie Beam Engine Trust; Rose Teanby; Geoff Wallis at Avon Industrial Buildings Trust; The Canal and River Trust; The Museum of Scottish Industrial Life, Coatbridge; Cefn Coed Colliery Museum, Crynant; The Scottish National Mining Museum, Newtongrange; Myerton Motor Museum, East Lothian; Museum of Science and Industry, Manchester; The Thinktank Museum, Birmingham; St. Vincent Street Church, Glasgow; Crofton Pumping Station, Wiltshire; Brunel's SS *Great Britain*; Fakenham Gasworks Museum; Knockando Woolmill, Speyside; 'Steam' the Museum of the Great Western Railway, Swindon; Didcot Railway Centre; Blists Hill Victorian Town, Ironbridge Gorge Museums; The National Trust; The National Waterways Museum, Ellesmere Port; The National Gas Archive; Tower Bridge, London; The Tank Museum, Bovington, Dorset; Kelham Island Museum, Sheffield; Wadworth Brewery, Devizes; Sumburgh Head Lighthouse & Marine Life Centre, Shetland; The staff and volunteers at Portsmouth Historic Dockyard;

Thanks to John Spear for his help and advice, and to Keith Whittles at Whittles Publishing Ltd for his enthusiasm for this project.

In memory of Kath Hannavy 1953–2024 whose support and encouragements made all my projects possible.

PREFACE

There is an enduring fascination in exploring the stories which underwrite Britain's industrial heritage. Some of the conversations I have had over the years while visiting and photographing surviving industrial sites have set me off on months of research, unravelling the relationships and links which connect pivotal figures in industrial history.

Getting a little closer to understanding the motivations of the visionaries who made it all happen is what continually piques my interest, and when I set out to start work on this project, I was briefly under the illusion that exploring the myriad links between the great engineers who built Britain was a novel – perhaps even original – idea. That belief was very quickly dispelled by the realisation that it was neither original nor novel. The eminent Victorian writer Samuel Smiles had beaten me to it by a hundred and fifty years.

In his day, Smiles was described as a 'Propagandist for Victorian values', and he virtually invented the genre of 'industrial biography', which enjoyed a considerable readership. His engineering biographies were ground-breaking and, written closer to the lives and events they chronicled, are highly illuminating. It has been fascinating reading his, and other, contemporary accounts of great industrial milestones, and realising that history has not always attributed the same importance to some inventors as writers predicted at the time.

But while he wrote at length about the same subjects as I, he did not have the advantages of either hindsight or modern colour photography with which to illuminate his biographies and bring their achievements to life.

However, the realisation that my idea was not original was a truth which chimed nicely with my central thesis in exploring the interactions, the cross-fertilisations and the collaborations which drove engineering and technological expertise forward – that the evolution of Britain's global industrial dominance was anything but linear, and, as in other walks of life, was certainly not realised by individuals working alone. The same names crop up again and again and, as can be read in the pages which follow, their influences on each other were both frequent and crucial. Many of them worked towards the same outcomes – either together, in parallel, or sequentially – and all left their marks on today's Britain.

In addition to Samuel Smiles, I am, as always, indebted to the many experts and volunteers at today's industrial heritage sites, museums and libraries, who have welcomed me, given me access to their collections, and shared their immense knowledge with me. Without them – and their willingness to let me roam freely with my camera – this book would not have been possible. My indebtedness to their input is acknowledged at the end of this book.

John Hannavy 2023

[opposite] The Dudley Canal and towpath, running through the Black Country Living Museum – built in the late 18th century by Thomas Dadford Jnr, using James Brindley's templates for narrow locks.

[below] Wendy Taylor's Grade II listed 'Timepiece Sundial' sculpture at the entrance to St. Katharine Dock in London reflects the proximity of the Greenwich Meridian. In 1971 when creating the sculpture, Taylor had her studio at St. Katharine Dock, long before Thomas Telford's eighteenth century dock and warehouses were restored as offices and apartments.

AN INDUSTRIAL TIMELINE

The engineers and inventors featured in this book span a period of two and a half centuries. This timeline seeks to provide a basic chronology while highlighting some of the links between 'The Men Who Invented Britain'

1708	• Abraham Darby introduces coke instead of charcoal when smelting iron.
1712	• Thomas Newcomen develops steam engine.
1715	• First Newcomen engines installed at Griff Colliery.
1724	• Daniel Defoe suggests a canal between the Forth and the Clyde.
1740	• Benjamin Huntsman develops a process for making steel.
1753	• John Smeaton elected Fellow of the Royal Society, aged 29.
1759	• Duke of Bridgewater commissions James Brindley to build Bridgewater Canal.
	• Smeaton's Eddystone Lighthouse completed.
1761	• Brindley's Bridgewater Canal opens.
1764/5	• James Hargreaves invents the Spinning Jenny.
1765	• James Watt develops the first steam engine with a separate condenser.
	• HMS *Victory* launched.
1766	• Matthew Boulton's Soho Manufactory established.
1768	• Smeaton starts work on Forth & Clyde Canal.
1769	• Watt surveys route for the Monkland Canal.
1770	• Work starts on Leeds to Liverpool Canal.
1771	• Richard Arkwright establishes the world's first water-powered spinning mill at Cromford.
	• Smeaton's Tay Bridge at Perth completed.
1772	• James Brindley dies.
1775	• Partnership of Boulton and Watt established and Soho Foundry opens.
	• Steam Engine Act passed by Parliament.
1776	• Trent & Mersey Canal opens.
1777	• Watt's 'Smethwick Engine' built
1779	• Samuel Crompton invents the Spinning Mule.
	• William Murdoch Patents 'D slide valve'.
	• Abraham Darby III builds the world's first Iron Bridge in Coalbrookdale.
1781	• Watt builds first engine fitted with William Murdoch's 'Sun and Planet' gears.
1783	• John Rennie meets Boulton and Watt at Soho Foundry, and is later employed by them.
1784	• Joseph Bramah patents 'unpickable' lock.
	• Murdoch builds steam-powered vehicle.

1786	• David Dale and Richard Arkwright establish their new mill at New Lanark on the Clyde.
	• Rennie works on Matthew Boulton's Albion Mill in Southwark.
1787	• John Wilkinson builds first iron-hulled boat.
1789	• Henry Maudslay works for Joseph Bramah.
	• William Jessop builds first railway using flanged wheels on iron rails.
1790	• Rennie appointed Surveyor of the Kennet and Avon Canal Company.
1791	• Brindley's Barton Aqueduct completed.
1792	• Smeaton's first enclosed harbour completed at Charlestown, Cornwall.
1793	• Hay Inclined Plane opens at Coalbrookdale.
1795	• Bramah patents his Hydraulic Press.
1797	• Maudslay sets up his own workshop.
1800	• Maudslay builds screw-cutting lathe.
	• Rennie appointed Project Engineer to develop Leith Docks.s
1801	• Marc Brunel patents block-making machinery.
	• Boulton and Watt introduce iron beams rather than wooden beams on to their engines.
	• William Murdoch appointed Chief Engineer at Boulton & Watt.
	• Rennie's Crinan Canal completed.
1802	• Matthew Murray builds his first hypocycloidal engine.
	• Work starts on William Jessop's 'Floating Harbour' in Bristol.
1803	• William Symington's stern-paddler *Charlotte Dundas* sails on Forth & Clyde Canal.
1804	• Portsmouth Block Mills open – the world's first 'production line' factory.
1805	• Murdoch installs gas lighting in a Salford cotton mill – and on the street outside.
1805	• Thomas Telford and William Jessop's Pontcysllte Aqueduct opened.
1807	• Frederick Winsor installs gas lighting on Pall Mall, London
1808	• Robert Stevenson appointed Chief Engineer of the Northern Lighthouse Trust.
	• Murdoch awarded the Royal Society's Rumford Gold Medal.
1809	• Matthew Boulton dies.
1811	• Stevenson's Bell Rock Lighthouse completed.
	• Samuel Clegg installs gas lighting in Stonyhurst College, the first school thus lit.
1812	• Henry Bell's paddle steamer *Comet* sails
1815	• George Stephenson and Humphrey Davy independently develop miners' safety lamps.
1816	• Telford's Craigellachie Bridge opens.
1817	• Rennie's Waterloo Bridge completed.
1818	• Marc Brunel patents his tunnelling Shield.
1819	• James Watt dies.
	• Boulton and Watt offer iron beams in sixteen different sizes.
1822	• Telford's Caledonian Canal completed.
1825	• William Fairbairn built the 'Lions of Catrine', the world's largest waterwheels.
	• Stockton and Darlington Railway opens – first in the world to use steam traction.
1826	• Telford's Menai Straits and Conwy Suspension Bridges both open.
1827	• Nasymth steam carriage operates on Queensferry Road near Edinburgh
1828	• James Nasmyth is apprenticed to Henry Maudslay
	• Thomas Telford and Philip Hardwick's St Katharine Dock, London, completed.
	• Robert Stephenson builds *Lancashire Witch*, the first locomotive with a multi-tube boiler.
1829	• Fairbairn builds cast-iron panel bridges for Liverpool & Manchester Railway.
	• Robert Stephenson's *Rocket* excels at Rainhill Trials on Liverpool and Manchester Railway.
1831	• Rennie's London Bridge opens.
1834	• John Scott Russell's steam carriage trialled.

1836	• Fairbairn opens Millwall shipyard, later sold to John Scott Russell.
1837	• Fairbairn patents steam riveter.
	• James Nasmyth opens Bridgewater Foundry in Patricroft near Manchester.
1838	• Isambard Kingdom Brunel's broad gauge Great Western Railway opens.
	• Nasmyth designs his steam hammer.
1839	• William Fairbairn & Co. build their first steam locomotive.
1841	• Isambard Kingdom Brunel's SS. *Great Britain* launch in Bristol.
	• Isambard Kingdom Brunel's completes Box Tunnel on the Great Western Railway.
1842	• Nasmyth patents his steam hammer.
1844	• William Fairbairn and John Hetherington patent the Lancashire Boiler.
	• John Scott Russell publishes first paper on the nature of bow waves.
1846	• William Armstrong opens Elswick Works.
1847	• Jesse Hartley's Royal Albert Dock, Liverpool, completed.
1849	• Robert Stephenson's Tubular Railway Bridge at Conwy opens.
	• Isambard Kingdom Brunel's Swivel Bridge at Bristol Harbour installed.
1850	• Fairbairn patents tubular crane.
1851	• The Great Exhibition at the Crystal Palace, Hyde Park, London.
1855	• William Armstrong develops his first breech-loading guns for the British Army.
1857	• Samuel Smiles publishes his first industrial biography – *The Life of George Stephenson*.
1858	• Isambard Kingdom Brunel's SS *Great Eastern* launched at Millwall.
1859	• Isambard Kingdom Brunel dies aged 53.
1860	• John Scott Russell establishes Institution of Naval Architects.
	• HMS *Warrior* built.
1864	• Brunel's Clifton Suspension Bridge completed by Sir John Hawkshaw and William Barlow.
1865	• Henry Bessemer develops his 'Converter' for making steel.
1869	• The tea clipper *Cutty Sark* built on the Clyde.
1871	• William Arrol & Company established.
1872	• William Froude builds his first test tank to measure bow waves and refine hull shape.
1879	• Joseph Wilson Swan demonstrates his electric light bulb at Newcastle Lit & Phil.
	• Armstrong's Cragside becomes first house in the world lit by electricity.
	• Thomas Bouch's Tay Bridge collapses.
1890	• Forth Railway Bridge opens.
1892	• Alexandre Tropenas patents developments to Bessemer's converter.
1894	• Armstrong develops Tower Bridge hydraulics.
1896	• Arrol-Johnson car company established.
1907	• Arrol builds first Giant Cantilevered Crane.
1908	• Huge Arrol Gantry built at Harland & Wolff's shipyard in Belfast.
	• Ernest Shackleton takes modified 1908 Arrol-Johnson car on Antarctic Expedition.
1911	• RMS *Titanic* launched from the Arrol Gantry.

INTRODUCTION

There are very few – if any – structures which better exemplify the emergence of Britain as an industrial nation than the Iron Bridge across the Severn in Coalbrookdale in Shropshire, built in 1781. It was built by Abraham Darby III and is the most iconic structure in what is now the Ironbridge Gorge UNESCO World Heritage Site. His bridge is what gave the township its name and, more importantly, signalled a massive social and industrial revolution.

One of the most important figures in that revolution was Abraham Darby, grandfather of the bridge builder, who had pioneered the use of coke-fired blast furnaces to smelt iron in Bristol in 1708. Coke was made by heating coal in an oxygen deprived atmosphere to drive off the gas – which was later used for lightng and heating. He moved to premises in Coalbrookdale in Shropshire in 1709 where there was a ready supply of the low-sulphur coal which was best suited to coking.

Using coke rather than charcoal, the heat in the furnace was much higher and the fuel burned for longer before collapsing, speeding up the smelting process and reducing the cost. The remains of his early blast furnace can still be seen in Coalbrookdale. Without his achievements, the pace of Britain's industrial revolution would have been different.

Without cast iron, Thomas Newcomen and James Watt could never have developed their steam engines – and well over 100 cast iron cylinders for Newcomen engines had been manufactured at Coalbrookdale alone by the middle of the 18th century.

Without steam engines, the massive surge in industrial output when compared with water power would not have happened. Without the replacement of wood with cast iron and wrought iron, reliable and accurate machines would have remained a dream. Without the ability to accurately machine iron, Richard Trevithick, the Stephensons and others could never have pioneered the railways, and Brunel could never have built his great iron steamships.

Thanks to the Darby family and those who followed in their footsteps, the 18th and 19th centuries created a second Iron Age which radically changed the world and was arguably more important than the first.

Cast iron and wrought iron – and once its production costs were reduced, steel – turned out to be the most versatile building blocks of the industrial revolution, leading to the machine age and mechanisation.

But irrespective of the potential of the great industrial innovations – such as steam engines, power looms and all the other attempts to mechanise routine tasks which were being proposed towards the end of the 18th century – none of them would have found widespread adoption as

[opposite] Coalbrookdale's 1781 iron bridge.

[above] Just 70 years after Darby's bridge, iron foundries were producing work of great sophistication – celebrated at the Great Exhibition of 1851. This example graces Alexander 'Greek' Thomson's St. Vincent Street Church in Glasgow, opened in 1859. The church's many different column capitols were all cast at Weir and McIlroy's Glasgow foundry, whose output ranged from finely detailed work to heavy ironwork for the construction and railway industries. The company worked extensively for Thomson who had earlier designed McElroy's 'Italian Villa' at Cove on the Clyde.

quickly as they did had it not been for the development of an effective transport infrastructure.

Despite the scale and enduring impact of their contributions to the development of the modern world, the story of Britain's emergence as the world's first major industrial society revolves around a relatively small number of individuals. Their names crop up repeatedly through the major inventions of the 18th, 19th and early 20th centuries. The circles in which they lived and worked overlapped, creating vital links and cross-fertilisation of ideas.

It all had to start somewhere, and given the importance of milling in the centuries before industrial mechanisation, it is no srprise that some of the most important men in every locality were the millwrights. They had to be able to employ a wide range of skills to ensure their mills operated efficiently when required.

Without their skills, quite simply, there would be no bread. It is not surprising, therefore, that so many of the great engineers who emerged in the mid- to late-eighteenth century had been trained as millwrights. Amongst those millwrights are James Brindley, John Smeaton, William Fairbairn, Thomas Telford, John Rennie, William Murdoch, and others, who all earned their places in the pantheon of engineering history.

These were some of the men who not only invented the machines which powered Britain, but also created the transport infrastructure along which raw materials and finished goods could be moved to their customers or exported across the world.

[far left] This mark on the wall at Stone Mill is all that identifies this space as the former engine house. The flywheel sat in a deep pit now filled in. The marks are said by some to have been caused by years of 'barring' – notching the flywheel round using a crowbar until the piston and crank were in the optimum starting position.

[left] Before conversion to steam, Stone Mill was powered by two large undershot waterwheels fed from the nearby River Biss.

The rapid expansion of the textile industry is an particular example – without steam engines to power the machinery, the major industrial centres of Britain would never have been developed, and mills would have continued to be dependent on fast-flowing rivers to drive their water wheels. But building steam engines in Smethwick, as James Watt and Matthew Boulton did, would have counted for little had there been no reliable means of transporting them around the country to the factories they were built to power.

If industrial output was the first real beneficiary of the steam engine, the steam engine was the first real beneficiary of the developing canal system. Without canals, engines could not have been shipped across Britain, and without steam pumping engines, canals would have run short of water as traffic increased.

Boulton & Watt had 'assemblers' based across the country whose job it was to organise the erection of the engines, and usually maintain them thereafter. For example, their agent in Wiltshire was one George Haden (1788–1856), who had been apprenticed at their Soho Foundry in Smethwick at the age of 15 before being entrusted – aged just 22 – with the sale and installation of engines first in the north-west of England and later across parts of the south-west.

In late 1814, shortly after being assigned to the south-west, he arranged his first sale – a ten horsepower engine with a $19\,^3/_4$ inch diameter cylinder and a 30 inch stroke – to Stone Mill, a textile factory in Trowbridge, Wiltshire, owned and operated by John and Thomas Clark. The mill had hitherto been water-powered via a channel from the nearby River Biss. Delivery times for big engines would have been lengthy as these were bespoke items, tailored to the requirements of each purchaser.

It is worth remembering that the speed of the world at that time was governed by the speed of a walking horse – and even heavy horses could only move relatively slowly when hauling a heavily-laden canal boat. The time-frame would be further extended by the need to design and construct engine houses for those mills converting to steam from water-power.

Today, the 120 miles between Smethwick and Trowbridge can be travelled in about two hours, but back in 1815, even delivering a letter over that sort of distance could have taken days over largely rubble-filled roads. The first liveried mail coach had only been introduced – between London and Bristol – in 1784, with uniformed postmen first being introduced in

[opposite top left] The giant Pit Wheel in John Rennie's Claverton Pumping Station – which lifts water 48 feet from the River Avon to the Kennet & Avon Canal – has 204 teeth made up of 408 carefully profiled oak blocks held in place with dowels. Cutting and profiling the teeth by hand was a regular job for the millwright. Today they are profiled using a specially-built machine. When the pump was running constantly, the teeth would rarely get the chance to dry out, and thus lasted for longer without splitting. As Claverton Pumping Station sits in a flood plain, the machinery is regularly under water, and thus forever covered with silt.

[opposite top right] Part of the wooden gearing in the restored Moulin de Cotentin in Normandy – already showing the signs of wear.

[opposite bottom] Wooden gearing in the restored Bursledon Windmill in Hampshire.

[right] The water-powered Stanley Mills on the banks of the fast-flowing River Tay in Perthshire – the Duke of Atholl's huge complex of late 18th and early 19th century cotton mills right in the heart of wool country. Richard Arkwright was one of those who helped establish the mill. The East Mill was added to spin and weave flax, but was gutted by fire in 1799. David Dale helped in its re-construction after the fire.

1793. So, communications between Boulton & Watt and their customers would not have been either quick or easy – and postal pricing was still dependent upon the distance the letter had to travel.

The universal penny post, with Rowland Hill's innovative 'Penny Black' postage stamp, was still more than a quarter of a century in the future and the railway age which would revolutionise the speed of transport between major centres had not yet dawned.

For the first steam engine to be installed in Brick Mill in Trowbridge – next door to Stone Mill – the design of both the engine and the engine-house was personally undertaken by James Watt Junior, son of the great steam pioneer. Records of the supply and installation of the Stone Mill engine tell us that the engine itself cost £873.11s.0d (£873.55). A further £25.10s (£25.50) was spent on building a stone foundation for it, and Haden's fee – his name is spelled 'Hayden' in Clarks' records – was £30. In addition, the sum of £9.2s.6d (£9.12) went on wages for the team who assisted him.

Whilst Boulton and Watt built engines, they did not deliver them, leaving that to be arranged by the buyer. The journey from the Soho Foundry for that first engine, in late 1814 or early 1815, cost £33.19s (£33.95) from loading on to barges on James Brindley's Birmingham Canal at Smethwick, to unloading at the quayside at Hilperton Marsh on the

[below left] 'Spinning Wool on Skye' from a photograph by the George Washington Wilson studio of Aberdeen. The photograph probably dates from the 1870s or 1880s.

[below right] David Dale and Robert Owen's pioneering industrial complex at New Lanark was initially water-powered from the River Clyde, but steam engines were later.

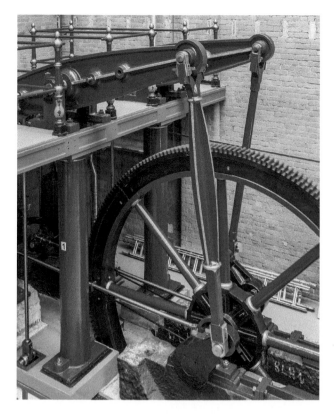

Kennet and Avon Canal – 5s.8d. (28p) per mile. For the last mile – from Hilperton Wharf to Stone Mill – local hauliers charged £6.16s.6d (£6.82). These were huge investments for the businesses which embraced the new technology, and they were repeated by the owners of thousands of mills and factories.

The textile industry into which these engines were being introduced was evolving rapidly – undergoing changes which were dominated by three men – James Hargreaves (1720–1778), Richard Arkwright (1732–1792) and Samuel Crompton (1753–1827). Their inventions revolutionised the manufacture of woven textiles, raising productivity and reducing prices.

Samuel Crompton was born in 1753 in a farm cottage at Firwood Fold, then a small group of properties just outside Bolton in Lancashire. He moved to a much more impressive address when the family rented the splendid early 16th century timbered Hall i'th'Wood in the town while he was still a young child.

Despite hard times hitting the family after his father's untimely death, Samuel's mother met the not insignificant cost of his education. From as early as he could cope with the spinning wheel, Samuel had been expected to contribute to the family income by spinning cotton, so it is hardly surprising that a Spinning Jenny – invented by James Hargreaves of Blackburn in the 1760s – was introduced into the Crompton household as early as 1769.

While being trained as both a spinner and a weaver, the young Crompton became disillusioned by the slow rate of production, the coarse quality, and lack of strength of spun cotton produced on the 'Jenny'. The mass production of textiles was never going to become possible with such a bottleneck at the very start of the process. The Jenny marked a significant step forward – before then, the spinning of yarn was achieved one thread at a time on a traditional spinning wheel, usually operated by women working at home.

[top left] This twin-beam engine, maker unknown, dates from 1840 and was operated in a Rochdale mill until 1953. It is now preserved in the Northern Mill Engine Society's Bolton Steam Museum.

[top right] No.9 Firwood Fold in Bolton where Samuel Crompton was born in 1753.

[above] Hall I'th'Wood, Bolton, which the family was renting at the time Crompton invented the spinning mule.

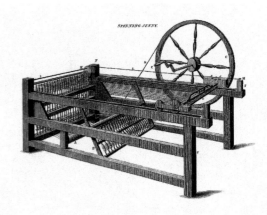

OLD SPINNING JENNY.

[top left] A contemporary illustration of Arkwright's Water Frame

[top right] A late 18th century illustration of Hargreaves' Spinning Jenny.

[above] This early 20th century postcard captioned 'An Old Spinning Jenny' in fact shows an early version of Crompton's Mule, but by that time it would appear that the terms 'spinning jenny', 'spinning mule' and 'mule jenny' were often used interchangeably. This machine is now in the collection of Bolton Museum and Art Gallery.

The name 'Spinning Jenny' has repeatedly been claimed to have been adopted by Hargreaves because that was the name of his wife or one of his daughters, but records show that none of them was named Jenny. More likely is the corruption of the words 'gin' or 'ginny', northern terms for an engine at that time.

A significant success in the mechanisation of cotton and wool spinning is credited to a third Lancastrian, Preston-born Richard (later Sir Richard) Arkwright who, in the 1760s, ran a barber's shop in Bolton's Churchgate where he was also renowned as a 'peruke-maker', a manufacturer of wigs. So it is to Arkwright and the invention of the 'Spinning Frame', also referred to as the 'Water Frame', that history has assigned the credit for initiating the industrialisation of textile manufacture.

Without him, the great mills of the 19th century would never have been needed, and the availability of cheap fabrics would never have come about. By the closing years of the 18th century, one spinner working in a factory could be in charge of a machine capable of spinning hundreds of threads at a time rather than the single thread on the home spinning wheel.

Arkwright's carding and spinning machines were first installed in his Derbyshire mill at Cromford in 1771, initially driven by horses, and at several other locations in which he had an interest – including Stanley Mills on the banks of the Tay near Dunkeld, where power was drawn from the fast-moving river.

He also had an early but relatively brief involvement in the design and development of the huge water-powered mill complex at New Lanark. David Dale, a Glasgow banker, and Arkwright formed a partnership in 1784, acquiring the site which they named New Lanark, and starting work on the mill and its model community. It was a short-lived partnership and they went their separate ways after just two years – even before the first mill on the site had been completed.

New Lanark's development was largely driven by William Kelly, appointed by Dale as General Manager. Kelly's 1792 Patent No.1879 *Certain New-Constructed Machinery, to be Applied to Spinning Machines for the Purpose of Conducting the Process of Spinning in a much more Expeditious and less Expensive Manner than by any Mode hitherto discovered* was intended 'to enable those machines, commonly known by the names of roving billies and slobbing and common and mule jennies, used in preparing and spinning cotton, wool, sheep's wool, flax, and other staple to be worked, both out and in, with a continued motion'.

However, according to contemporary records, the mills were initially equipped with 'Spinning Jennies'. Kelly also played a significant role in their partial automation using what were described as 'patent jennies' as opposed to manually-operated 'common jennies'. The conclusion to his patent summarised the 'improvements:

> The above is a particular description of the nature and use of those new invented improvements in spinning machinery, by the application of which all and every kind of these spinning machines, commonly called jennies, may be spun by water, by steam, or by horses instead of manual labour only, which has been the practice hitherto, and these improvements are applicable not only to machines which may be made in future, but also to those at present in use, which will be of great public utility.'

However more than 12 years before that, Samuel Crompton's experiments in search of a better way of spinning had led him to design what he called the 'Hall i'th'Wood Wheel' or the 'Muslin Wheel' after a series of experiments between 1774 and 1779. It would be developed into the most successful and enduring of the three developments which revolutionised cotton production throughout the world. His 'Muslin Wheel' later became known as the 'Spinning Mule' some say because it was a hybrid which combined the best elements of both Hargreaves' and Arkwright's inventions. While it was Arkwright's frame which initially drove the mechanisation of spinning, Lancashire favoured the evolutions of Crompton's Mule.

With both of these machines, cotton could be spun more quickly, more consistently, and to finer gauges than ever before. Thanks to the spinning mule and the spinning frame, Lancashire and Cheshire weavers had finer threads to work with, and could produce higher quality woven cotton cloths, as well as the utilitarian 'fustian' – a thick twilled fabric – which they sold across the world. However, the massive growth in textile manufacture brought about in the early decades of the machine age would never have happened without the many pioneering inventors and builders who helped develop the essential infrastructure for an emerging industrial nation.

[top] An 1891 Cornish boiler by little-known maker William Lord of Lord Street Iron Works, Bury. It raised steam for the single-cylinder 'nodding donkey' pump at Lion Salt Works in Cheshire.

[above] The single fire tube 'Cornish' boilers at Bursledon Brickworks in Hampshire raised steam for a horizontal single-cylinder engine by John Wood of Wigan. Via belt drive, huge cogs and complex gearing, it drove a Bennett & Sayer brickmaking machine – engineering on a grand scale, which it had to be in order to work with the heavy wet clay.

Inventions do not happen in a vacuum, nor are they usually the work of just one man. The best emerge after a sharing of ideas – intentionally or unintentionally – and a shared belief amongst inventors that each has that essential insight into how a task can be performed more effectively, more quickly, or at a lower cost.

In that respect Henry Bessemer erroneously believed himself to be unique 'inasmuch as I had no fixed ideas derived from long-established practice to control and bias my mind', but every other inventor would believe the same of himself.

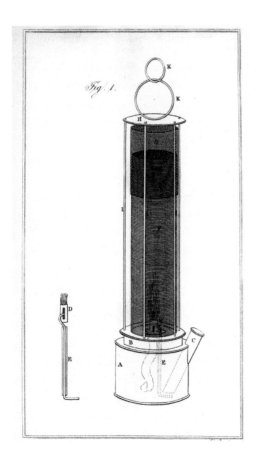

[above] George Stephenson's safety lamp, published in 1817 in support of his claim that the announcement of his lamp pre-dated Davy's. While his lamp – known locally as the 'Geordie Lamp' and widely used in the north-east – was both first and brighter, Davy's was ultimately proved to be safer.

Through both the scientific societies, and Literary & Philosophical Societies which were established in the major centres across the country, many of the great pioneers met, discussed their work, shared ideas, and offered each other suggestions and insights which helped propel science and engineering forwards. These societies became melting pots for innovative thinking across the widest range of disciplines. Before the advent of rapid mass communications, meeting face-to-face to discuss the issues of the day was an essential catalyst for industrial progress.

Literary & Philosophical Societies were established in a number of major towns and cities in Britain in the second half of the 18th century and into the 19th, initially as discussion groups, but their members' wide-ranging interests made them evolve into much more. In the late 18th century, the word 'Philosophical' was a synonym for 'Scientific', so these new societies embraced the sciences, literature and the arts. By the early 19th century they had become magnets for inventors, pioneers and visionaries across a wide spectrum of disciplines and a forum for discussing the ideas which fuelled the industrial revolution.

When the Leeds society was established in 1819, it actually styled itself the Leeds Philosophical & Literary Society, but elsewhere, the term 'Lit and Phil' became a widely used abbreviation. The Newcastle 'Lit & Phil', established in 1793, was where the railway pioneer George Stephenson demonstrated his prototype miner's safety lamp in 1815 – the same year that Humphrey Davy demonstrated his safety lamp in London.

The two men had, independently, arrived at very similar solutions to the perennial problem of dealing with fire-damp – the explosive mixture of methane and other gases which gathered deep underground in collieries – although Stephenson was surprisingly very vocal in the accusations of plagiarism which he levelled at Davy.

Someone who would have otherwise been keen to have Stephenson's views on the design of railway locomotives was fellow member John Buddle who, with William Chapman, is credited with the development – also in 1815 – of the six-wheeled locomotive *Steam Elephant*, a replica of which can sometimes be seen in steam at Beamish Museum. But Buddle and Stephenson were on opposite sides of the safety lamp furore, with Biddle supporting the priority of Davy's design.

Joseph Wilson Swan demonstrated his electric light bulb by illuminating one of the Newcastle Lit & Phil's rooms with it for the first time in 1879. It can be no surprise, therefore that another member, William Armstrong, commissioned Swan to install his electric light bulbs at his 'hunting lodge' which he was enlarging and modernising at Cragside in Northumberland. That same year, Cragside became the first private house in the world to be lit by Swan lamps.

The Manchester Literary & Philosophical Society, established in 1781, is the second oldest learned society in the country after the Royal Society of London for Improving Natural Knowledge founded in 1660 – now known simply as The Royal Society.

The Manchester society's eminent 18th and 19th century members included the chemist John Dalton (1766–1844), James Prescott Joule (1818–1889) after whom the SI unit of energy is named, and the engineer William Fairbairn (1789–1874). Other members included the engineer Joseph Whitworth (1803–1887), chemist, cotton tycoon and pioneer photographer Joseph Sidebotham (1824–1885), and inventor of the steam hammer James Nasmyth.

The idea that the likes of Joule, Nasmyth, Sidebotham, Fairbairn and Whitworth could have been in the same room together and engaged in a discussion about the emerging technologies of the day is a tantalising one. These men were amongst the engineering and scientific titans of their time, responsible for many of the major developments in 19th century industry, and the cross-fertilisation of their ideas proved hugely important.

Whitworth, following on from the pioneering work of Henry Maudslay, was responsible for the simple but hugely important standardisation of screw threads. Before he developed the Whitworth Standard, engineers had to match the thread of a nut to the thread of its partner bolt, and as they were all cut by hand, accuracy was not assured. Whitworth's measuring equipment was central to the drive towards precision, and the birth of the science of metrology – the study of measurement. His machines for manufacturing and threading bolts were crucial in enabling industrial mass production and repeatability without sacrificing accuracy. The quest for precision had driven Whitworth for most of his life, from his earliest days learning his craft in London under the finest engineers of his day – including Henry Maudslay, the father of the precision machine tool – to heading one of the century's largest industrial companies.

To get to London as a 22-year-old, he is said to have travelled on a working canal boat. Along the way he met, fell for and later married, Fanny Ankers, the daughter of a successful owner of canal boats in Cheshire. In later years, his Manchester operations would merge with Sir W. G. Armstrong & Company to create the Armstrong-Whitworth conglomerate whose catalogue of products would be very wide-ranging indeed and included armaments, ships, locomotives, cars, and aircraft.

Long before that, however, the evolving steam engine moved from being a simple low-pressure machine to using high pressure steam and requiring expert management. The era of the mechanical engineer dawned. The stoker went from being a navvy with a shovel on the likes of Brunel's SS *Great Britain*, to being a skilled fireman, and when in charge of the furnace on a mill engine, this skill was every bit as important in the efficient operation of the machine as was the engineman's. The bigger the engine, the greater precision needed to ensure the fire burned evenly across the entire grate area. An uneven fire could cause variations in steam pressure – and that would cause inconsistencies in output.

All those innovations, brought together in a boiler designed to increasingly precise specifications and manufactured to exacting standards made William Fairbairn's Lancashire boiler the most efficient of its day.

Development did not stand still – other boiler-makers would improve the design incrementally over the following decades so that by the 1870s when boiler-makers had started to move from using wrought iron to Bessemer steel, the risk of boilers exploding under pressure had been significantly reduced whilst those operating pressures had themselves been greatly increased.

[top] Alan Keef's 2002-built replica of Buddle and Chapman's *Steam Elephant*, seen here on the Pockerley Waggonway at Beamish Museum in Northumberland. The original initially performed poorly, but things improved when the original wooden rails on which it ran were replaced with iron ones.

[above] A horse-drawn narrowboat on the Cromford Canal at Whatstandwell in Derbyshire, c.1904. William Jessop worked with Benjamin Outram on the 14-mile canal from Cromford to Langley Mill, which required 14 locks and the excavation of four tunnels. Work started in 1789 and was completed in 1794.

By the time the great cotton mills were being built – with their large high-pressure triple-expansion engines – the Lancashire boiler was a very efficient design indeed. One of the most successful modifications to the basic design was introduced and patented in 1851 by W. & J. Galloway, the great Manchester engine-builders. They took Fairbairn's ideas a step further. As maximising the area of the fire/water interface was the key to getting maximum benefit from the fire and maximum heat transfer to the water, wide diameter 'Galloway Tubes' were built into the back half of the fire tubes, increasing that surface area considerably. The idea was adopted by a number of other manufacturers and soon became a standard feature, continuing in use until the end of the steam era.

Lancashire boilers came in a range of sizes, usually between 18ft and 30ft (5.5 to 9 metres) long, and with a diameter between 5ft 6ins feet and 7ft 6ins (1.6 to 2.3 metres). The boilers at Crofton Pumping Station in Wiltshire – at 7ft 6ins (2.3 m) in diameter and 28ft (8.5m) long – are typical of those most widely installed. Boilers of that size have a water capacity of around 4,000 gallons (18,000 litres).

From the installation of the first Lancashire boiler – probably around 1845 – to the last being withdrawn in the 1960s, this efficient design remained in everyday use for well over a century. Of course the last to be installed was a much more sophisticated and efficient device than the earliest, but that's progress. At the same time, with the increasing availability and quality of cast and wrought iron, machinery moved from being built of wood, which lacked both rigidity and durability, to being built of iron.

A number of early engineers took their knowledge into other enterprises, expanding their skills in the forging and shaping of iron into boat-building, railway engineering and the wide range of other industrial developments in which Britain would lead the world. Their achievements, and those pioneers who preceded them, would later be celebrated in the writings of Samuel Smiles.

The first iron-hulled boat is believed to have been built by John Wilkinson at his Bradley Forge back in 1787 and was used on the Severn.

After one engineer, Henry Creighton, left Boulton and Watt's employment at the Smethwick foundry, he was involved in the design of a number of iron-hulled boats for the Forth and Clyde Navigation in Scotland. Creighton and his brother William were both one-time agents for the company, and the first of Henry's boats, *Vulcan*, was the first iron passenger craft operating on the Forth and Clyde Canal. Based on a hull shape by naval architect Admiral John Schank, *Vulcan* was built at Thomas Wilson's Faskine boatyard using iron plates rolled at the Calderbank Works of the Monkland Steel and Iron Company. A 1988 replica – the original was completed in 1819 – is displayed in Coatbridge at the Summerlee Museum of Scottish Industrial Life. It could carry more than 200 people, some inside the saloon, others on an open top deck. It is believed to have been the first riveted iron-hulled passenger-carrying vessel in the world, and one of the earliest iron boats ever built. It was, of course, horse-drawn, and because of its size and laden weight it have required a team of horses.

[below] Blaenavon Ironworks cast houses with the remains of one of the original 1787 blast furnaces behind the right hand cast house. Molten low grade iron flowed along a branched channel from the furnace into pig beds on the cast house floor, creating a cast which looked like piglets being suckled by their mother – hence 'pig iron'.

[bottom] The replica of the 1819 horse-drawn barge *Vulcan* sits over the Monkland Canal at the Museum of Scottish Industrial Life. It was built of riveted mild steel at the Linthouse Yard on the Clyde, for the 1988 Glasgow Garden Festival.

By the middle of the 19th century it must have been hard to recall a world without wrought and cast iron, so ubiquitous were the materials. The catalogue of the 1851 Great Exhibition in Hyde Park bears testament to the range and quality of iron products available – and to the intricate detail of many of the mouldings being produced. Working with cast iron had been transformed from a simple affair to a very highly developed skill. To facilitate that evolution, the quality of the basic materials had been refined, casting methods had been improved, and a whole range of specialist tools had been invented. Within a relatively short time frame, the 'engineer' had become a specialist, his palette of skills dramatically expanded.

Much more cost-effective methods of producing high quality steel had been developed, which added yet another affordable material to the designer's and engineer's choices, and also expanded the range of essential skills which had to be learned and exploited.

Keeping abreast of all those developments must have presented a great many challenges to those seeking to progress in this new industrial world. The theories, techniques and processes we now take for granted were evolved during those years, and refined by engineers and inventors who did it all without the benefits of computers or the networking solutions we all rely on today, and yet successfully pioneered the quest for precision.

The prospect of today's computers was hinted at in the work of Charles Babbage (1791–1871) in his ideas first for a 'Difference Engine' and subsequently for an 'Analytical Engine'. However, it was the mathematical prowess of Lord Byron's daughter Ada Lovelace (1815–1852) which would be recognised by history as having created the first algorithm specifically written for a computing machine. Babbage's machine, however, was not built or put to the test during his lifetime.

[above left] The casting floor at Dinorwig Foundry, now the Welsh Slate Museum, a UNESCO World Heritage Site. In its heyday, the workers could repair, rebuild or construct new, every item of equipment needed to keep the quarries and their railways operational. The machinery was driven by line-shafting from a huge waterwheel.

[above] Matthew Murray's compact 5hp (3.5kW) hypocycloidal engine, built at his Round Foundry in Holbeck, Leeds in 1802, delivered more torque than other similarly-sized engines. It used a novel way of converting reciprocal motion into rotative motion. Installed new in John Bradley & Company in Stourbridge, it may have driven a small rolling mill. One of two surviving examples, it is now in Birmingham's Thinktank Museum. The other is in the Henry Ford Museum in Dearborn, USA.

SAMUEL SMILES AND *LIVES OF THE ENGINEERS*

At least a century before the present writer had the idea of exploring the threads which linked their lives and achievements, those same engineering visionaries had attracted the attention of Samuel Smiles – one-time Editor of the *Leeds Times* newspaper and the prolific author of books with such prosaic titles as *Self-Help, Character, Thrift* and *Duty* – inspired by series of lectures he had delivered to Leeds working men's groups in the 1840s. Through his books he was described as a 'propagandist for Victorian values', but he was equally in admiration of those responsible for great engineering milestones in the 17th and 18th centuries.

Smiles was born in the East Lothian town of Haddington, and initially set out to become a doctor having signed up for a medical apprenticeship before enrolling at Edinburgh University at the age of 17. There his interests turned to politics and social reform.

He was the first to write accessible histories of those he believed had made the most important contributions to Britain's industrial development, including James Brindley, Thomas Telford, John Smeaton, John Rennie, Robert and George Stephenson, Matthew Boulton and James Watt, and James Nasmyth. His five-volume *Lives of the Engineers* is recognised as a milestone in the published history of those who 'invented' Industrial Britain. Before that series was published in 1862 he had written *The Life of George Stephenson* (1857), and in 1858 he published *Industrial Biography: Iron workers and toolmakers* which explored the lives of many other industrial pioneers such as William Fairbairn and Henry Maudslay. A new edition of *Lives* was published in 1874, and in 1884 he further developed his biographical theme with *Men of Invention and Industry*.

In 1885 he edited and published *James Nasmyth, engineer, an autobiography* and, in 1894, came *Josiah Wedgwood, his Personal History*. By the time of his death, he had established himself as its primary exponent. In the introduction to *Industrial Biography* in 1863, he had explained that

> 'Without exaggerating the importance of this class of biography, it may at least be averred that it has not yet received its due share of attention. While commemorating the labours and honouring the names of those who have striven to elevate man above the material and mechanical, the labours of the important industrial class to whom society owes so much of its comfort and well-being are also entitled to consideration.'

A caricature of Samuel Smiles, one of one thousand three hundred and twenty-five cartoons produced for *Vanity Fair* between 1873 and 1911 by Sir Leslie Matthew Ward (1851–1922) under his pen name 'Spy'. Ward is reported to have been paid between £300 and £400 pounds for each portrait. This cartoon was published in January 1892 celebrating Smiles's eightieth birthday.

opposite: Lime Kilns on a spur of the Dudley Canal. Smiles was fascinated by the canal-builders and their lock systems, their building methods and materials.

[top] Briggate in Leeds, where the offices of the Leeds News were located. From a lantern slide by James Valentine of Dundee. Smiles edited the newspaper from 1839 until 1848.

[above] The River North Tyne which flows through Haddington where Smiles was born in December 1812.

Smiles was the author of many memorable quotes, addressing his Victorian audiences of working men in language to which they could readily relate, and in 1859 he had written

'The most important results in daily life are to be obtained, not through the exercise of extraordinary powers, such as genius and intellect, but through the energetic use of simple means and ordinary qualities, with which nearly all human individuals have been more or less endowed.'

While he was a prolific and sometimes overly opinionated biographer of others, he sought to convey the idea that he was the modest and somewhat reluctant author of his own autobiography, writing in the introduction

'Mr. Haigh was a very intelligent man, and a great reader, especially of biography. Many years since, he asked me, "Have you written your Autobiography yet?"

"Oh, no!" I answered, "there is no probability of that ever being done. I am too busy, besides, with other things that I wish to finish. I have been interviewed, it is true, like most other book writers, artists and men of notoriety. But my life has been comparatively uneventful; there is really nothing in it."

"Nothing in it?" responded my Mentor. "Why, your books are extensively read in this country and America, they have been translated into nearly every language in Europe. They appear in many of the Indian languages, and even in Siamese and Japanese. I am quite sure that your readers would like to know much more about yourself than has yet been published by your interviewers."

"That may be," I said, "but I do not think there are any passages in my life likely to be interesting to the public. My books, such as they are, must speak for themselves, without any biographic introduction."

"Well!" he observed finally, "think of my advice; I am persuaded that a history of yourself would be more interesting than any of your books."

Suitably provoked, Smiles eventually did start work on his autobiography, but it would not be published until 1905, the year after his death in April 1904 at the age of 91. It transpired that Mr. Haigh was correct – many people did buy and read *The Autobiography of Samuel Smiles, LL.D.* when it was published by John Murray.

Those he had chosen as subjects for his '*Lives*' were selected for the importance of their contribution and, in several cases, probably also for having also journeyed from humble beginnings to the pinnacle of success. Such characteristics fitted in well with his deeply-held thesis that any man can achieve great things by honest toil and integrity.

In that respect, some of the later figures featured on these pages might not have fitted his criteria, coming, as they did, from somewhat better-educated and wealthier backgrounds than many of those whose achievements he chronicled in great detail.But several of those whose achievements he championed existed in the shadow of their more illustrious masters.

In *Industrial Biography*, for example, he devoted a great deal of space to the story of the Darby family, and the rise of Coalbrookdale in importance, especially celebrating the work of the Darbys' business partner and forge-master, Richard Reynolds, some of whose innovations were evolved from the Cranege brothers' improved methods of refining iron, but his other innovations were to have equally widespread and enduring impacts. Smiles clearly thought Reynolds' contribution to industrial progress had been insufficiently celebrated. In *Industrial Biography* he wrote:

> 'Among the important improvements introduced by Mr. Reynolds while managing the Coalbrookdale Works, was the adoption by him for the first time of iron instead of wooden rails in the tram-roads along which coal and iron were conveyed from one part of the works to another, as well as to the loading-places along the river Severn. He observed that the wooden rails soon became decayed, besides being liable to be broken by the heavy loads passing over them, occasioning much loss of time, interruption to business, and heavy expenses in repairs. It occurred to him that these inconveniences would be obviated by the use of rails of cast-iron; and, having tried an experiment with them, it answered so well, that in 1767 the whole of the wooden rails were taken up and replaced by rails of iron. Thus was the era of iron railroads fairly initiated at Coalbrookdale, and the example of Mr. Reynolds was shortly after followed on all the tramroads throughout the Country.'

Thus were some of the the early 'foundations' laid for the future development of the railways and tramways which eventually brought about massive changes in the transportation and availability of raw materials and manufactured goods across the country in the nineteenth century.

Smiles was clearly very determined that Reynolds' place in history should be celebrated, albeit a century after the events which brought him recognition. Describing the rapid expansion of the works at Coalbrookdale, and the growth of the iron industry, he wrote:

> 'In 1784, when the government of the day proposed to levy a tax on pit-coal, Richard Reynolds strongly urged upon Mr. Pitt, then Chancellor of the Exchequer, as well as on Lord Gower, afterwards Marquis of Stafford, the impolicy of such a tax. To the latter he represented that large capitals had been invested in the iron trade, which was with difficulty carried on in the face of the competition with Swedish and Russian iron. At Coalbrookdale, sixteen "fire engines," as steam engines were first called, were then at work, eight blast-furnaces and nine forges, besides the air furnaces and mills at the foundry, which, with the levels, roads, and more than twenty miles of iron railways, gave employment to a very large number of people. "The advancement of the iron trade within these few years," said he, "has been prodigious. It was thought, and justly, that the making of pig-iron with pit coal was a great acquisition to the country by saving the wood and supplying a material to manufacturers, the production of which, by the consumption of all the wood the country produced, was formerly unequal to the demand, and the nail trade, perhaps the most considerable of any one article of manufactured iron, would have been lost to this country had it not been found practicable to make nails of iron made with pit coal.'

Clearly, Smiles believed that the emphasis on the work of Abraham Darby III in the development of the iron bridge had resulted in Reynolds' input appearing to be of lesser importance, and he was determined to rectify that, adding 'We need scarcely add that the subsequent history of the iron trade abundantly justified the sagacious anticipations of Richard Reynolds.'

When Reynolds retired, he passed his business interests on to his two sons, one of whom, William, became a major figure in his own right – Smiles was as fulsome in his praise of William's contributions to engineering as he had been writing about his father.

Summerlee Iron Works, Coatbridge.. RELIABLE SERIES. R1848

A postcard of the giant Summerlee Ironworks and Monkland Canal in Coatbridge, Lanarkshire, around 1906. It was one of the first sites in Britain to use the innovative 'hot blast' process for making iron pioneered by James Beaumont Neilson, but it fell victim to the great depression in the 1930s and was demolished. Today the world's only surviving example of a hot blast iron furnace is in Germany. The Summerlee site is now home to the Museum of Scottish Industrial Life. The site contains a surviving fragment of a spur run from the Monklands Canal to service nineteenth century ironworks. The main route of the canal had been surveyed by James Watt. Ongoing archaeological excavations are revealing tantalising traces of Summerlee's industrial past. Smiles' *Lives of Boulton and Watt* was published in 1865 by London John Murray, and in Philadelphia by Lippincott & Company.

Other major influencers in the evolution of the iron industry were given similar detailed treatment – amongst them Henry Cort (c.1740–1800), whose development of the 'reverberatory furnace' pioneered by the Cranege Brothers, and Peter Onions' puddling process, improved the quality of iron. Smiles wrote that

> 'As early as 1786, Lord Sheffield recognised the great national importance of Cort's improvements in the following words:—"If Mr. Cort's very ingenious and meritorious improvements in the art of making and working iron, the steam-engine of Boulton and Watt, and Lord Dundonald's discovery of making coke at half the present price, should all succeed, it is not asserting too much to say that the result will be more advantageous to Great Britain than the possession of the thirteen colonies (of America); for it will give the complete command of the iron trade to this country, with its vast advantages to navigation." It is scarcely necessary here to point out how completely the anticipations of Lord Sheffield have been fulfilled, sanguine though they might appear to be when uttered some seventy-six years ago.'

He also championed many other figures from the early days of industrial iron-making, devoting an entire chapter to the importance of James Beaumont Neilson (1792–1865) whose 'hot blast' process for iron-making enabled costs to be reduced by using coal rather than coke in the smelting process. The huge Summerlee Ironworks in Coatbridge – eight miles east of Glasgow and just a few miles from Neilson's birthplace at Shettleston – was one of the first to adopt his ideas, and by 1857 it already had a workforce of more than 400.

The locksmith Joseph Bramah and his one-time works manager and pioneering engineer Henry Maudslay, Marc Brunel father of Isambard, and many others, were all subjects for Smiles' pen.

In the last of his 'biographies', *Men of Invention and Industry* published in 1884, he turned his attention to the world of shipbuilding, which had undergone rapid and massive change during the 19th century.

The pioneering work of Henry Bell in commissioning the building of the first passenger-carrying steamboat in Britain, the paddle-steamer *Comet*, is covered only briefly, but the work of Francis Pettit Smith in predicting the practical dominance of the screw propeller in ship design was explored in detail. Smith was responsible for the construction of the first steamer to be fitted with a propeller – the SS *Archimedes* built for the Screw Propeller Company at Ratcliffe Cross Dock on the Thames and completed in 1839.

Whilst he did not invent the propeller – several others were independently developing it as an alternative to the paddle-wheel – Smith's patent defined the optimum position of the propeller between the stern and the rudder – the configuration which is still used today.

Several other inventors had suggested it be positioned aft of the rudder, but Smith's innovation not only used an otherwise empty space beneath the vessel's stern, it also considerably improved the steering of the ship by directing the fast-moving water from the screw past the rudder blade, thereby considerably improving its turning efficiency.

Amongst the earliest to show interest in Smith's invention were Messrs George and John Rennie, who had taken over the business interests of their father, John Rennie the Elder, and were developing a growing reputation for themselves as marine engineers. Their interest would lead them to become investors in The Screw Propeller Company.

As late as the 1840s, the paddle-box was still considered to be the default method of propelling steamboats. Isambard Kingdom Brunel originally designed his SS *Great Britain* as a paddle-steamer, but after reportedly borrowing and testing the SS *Archimedes* for several weeks, was so persuaded by the benefits of screw propulsion that part way through construction he modified design of his vessel and its engine to incorporate a four-bladed propeller designed by Smith.

[top] The partially exposed remains of the Summerlee Iron Works were excavated during the 1980s. The remainder lies beneath the main museum buildings. The Monkland Canal largely fell out of use in the late 19th century, when the adjacent Caledonian Railway opened.

[bottom] 'Herringbone' gearing on a steam-powered winch, early 1900s, in the collection of the Scottish Museum of Industrial Life

While others had decided there was no future to the screw propeller, Smith first demonstrated the practicality of his version in 1837, so it is remarkable that Brunel adopted the screw just three years later. Despite the propeller becoming the most effective means of moving a ship, Smith made nothing out of his foresight and died in relative poverty. His original propeller, however, actually did not function very well, and was replaced on *Archimedes* by a design patented, also in 1839, by James Lowe. During the period when the vessel was on loan to Brunel, several other designs were experimented with, before a six-bladed version was chosen for SS *Great Britain*.

Despite his undoubted interest in the development of steamships, surprisingly absent from Smiles' oeuvre, however, was a biography of Isambard Kingdom Brunel – without doubt one of the greatest Victorian engineering visionaries of them all – who had died in 1859.

[right] The simple paddles on the replica of Henry Bell's 1812-built steamer PS *Comet*, the first passenger-carrying steamer in Britain, now displayed in Port Glasgow. The vessel was only in service until 1820.

[far right] The replica propeller fitted to the restored SS *Great Britain* in its dry-dock in Bristol. When launched, the vessel was fitted with a six-bladed propeller, but that was replaced for a time by a four-bladed version designed by Brunel himself in the belief that it would make for smoother motion. The original six-bladed version was later refitted.

However, Smiles was not one for re-inventing the wheel, and he was aware that Brunel's son Isambard had already published a comprehensive biography of his father in 1870. Indeed, he referenced that biography in *Men of Invention and Industry*.

From the large to the small, Smiles then went on to chronicle the story of John Harrison's marine chronometer which introduced springs in place of the pendulum, which gave timepieces portability and brought hitherto undreamed of accuracy to the science of navigation.

Smiles was acutely aware that the histories he was assembling might distort the relative importance of those whose achievements he was seeking to place in context. In *Men of Invention and Industry* he lamented the number of key inventions the originators of which are not known:

> 'Some of the most valuable inventions have descended to us without the names of their authors having been preserved. We are the inheritors of an immense legacy of the results of labour and ingenuity, but we know not the names of our benefactors. Who invented the watch as a measure of time? Who invented the fast and loose pulley? Who invented the eccentric? Who, asks a mechanical enquirer, "invented the method of cutting screws with stocks and dies? Whoever he might be, he was certainly a great benefactor of his species. Yet (adds the writer) his name is not known, though the invention has been so recent."'

He also reflected on the inadequacies of the Victorian Patent system, whereby the granting of patents might be assumed to reflect originality of thinking which might not, in fact, have been attributable to just one individual. Confusions arising out of contradictory patent claims kept Victorian courts quite busy as actions and counter-actions were brought by inventors who saw others profit from their ideas.

The parallel and individual 'invention' of the safety lamp by Humphrey Davy and George Stephenson, mentioned earlier in this book, was just one of the examples he cited, while the development of the planing machine could, he said, be attributed to at least six people working quite independently of each other.

Holding Henry Maudslay up as a paradigm, while noting that he 'troubled himself very little about patenting his inventions', Smiles noted that while [Maudslay]

> 'had much toleration for modest and meritorious inventors, he had a great dislike for secret-mongers – schemers of the close cunning sort – and usually made short work of them. He had almost equal aversion for what he called the "fiddle-faddle inventors", with their omnibus patents, into which they packed every possible thing that their noddles could imagine.'

In that, he was ahead of his time, but it would be several decades before patent law was reformed to eliminate the patenting of untried ideas – and the assertion of patent claims over ideas long in the public domain. But Smiles was certain that Maudslay's importance could not be overstated

> 'The vigilant, the critical, and yet withal the generous eye of the master being over all his workmen, it will readily be understood how Maudslay's works came to be regarded as a first-class school for mechanical engineers. Every one felt that the quality of his workmanship was fully understood... Nor can Oxford and Cambridge men be prouder of the connection with their respective colleges than mechanics such as Whitworth, Nasmyth, Roberts, Muir, and Lewis, are of their connection with the school of Maudslay. For all these distinguished engineers at one time or another formed part of his working staff, and were trained to the exercise of their special abilities under his own eye.'

Interestingly, at a meeting of the Institution of Civil Engineers in 1858, Mr. Henry Maudslay reported that

> 'Nearly forty years ago, his grandfather had invented and worked some ingenious machinery for punching boiler-plates, for making iron water-tanks for the navy.'

Industrial progress, Smiles recognised, was the result of the often-independent threads of ideas being brought together and woven into outcomes which greatly exceeded the sum of their parts and resulted in the key developments and machine tools which drove industry forwards.

> 'But most mechanical inventions are of a very composite character, and are led up to by the labour and the study of a long succession of workers. Thus Savary and Newcomen led up to Watt; Cugnor, Murdock, and Tevithick to the Stephensons; and Maudslay to Clement, Roberts, Nasmyth, Whitworth, and many more mechanical inventors. There is scarcely a process in the arts but has in like manner engaged mind after mind in bringing it to perfection. "There is nothing", says Mr. Hawkshaw, "really worth having that man has obtained, that has not been the result of a combined and gradual process of investigation. A gifted individual comes across some old footmark, stumbles on a chain of previous research and inquiry. He meets, for instance, with a machine, the result of much previous labour; he modifies it, pulls it to pieces, constructs and reconstructs it, and by further trial and experiment he arrives at the long sought-for result."

In the 120 years since Smiles died, ongoing exploration of industrial history has raised the profiles of many whom he championed, while others slipped into obscurity.

JAMES BRINDLEY, SAILING OVER HILLS

Britain's first modern canals were designed and built by James Brindley (1716–1772). Other engineers would develop his ideas and build bigger and more complex waterways, but they all owed a considerable debt to this man, born in Tunstead, Derbyshire and who had initially trained as a millwright.

He was fascinated by water power, building several mills, and developing pumping systems for coal and other mines. Cheddleton Flint Mill's North Mill in the Churnet Valley in Staffordshire is believed to have been designed by Brindley – the mill's purpose being to grind flint into the fine power used in the manufacture of durable domestic ceramics by his friend Josiah Wedgwood.

One of his last commissions was to conduct a survey of a route for a canal spur past Cheddleton Mill to join the Caldon Canal, which would give direct access to the heart of the ceramics industry in Etruria, Stoke-on-Trent.

Brindley was a visionary, and his canals were revolutionary and the answer to a growing demand. But there were few initially who were willing to back his ideas financially. After all, he

[above] A contemporary portrait of James Brindley.

[left] The North Mill at Cheddleton. Flint Mill – on the left of the picture was designed by Brindley, specifically for the grinding of flint. The South Mill, to the right, was converted from a corn mill, while in the foreground two kilns used to dry the powered flint survive. The Mill sits by the Caldon Canal – surveying its route was Brindley's last project, and the chill he caught there is said to have hastened his death.

[opposite page] The stern and tiller of a brightly painted narrowboat, built to fit Brindley's narrow locks.

connections with:

- John Smeaton
- James Watt
- John Rennie
- Thomas Telford

was not an engineer, and had no prior experience of realising the network of waterways which he envisaged. But he was, as became apparent, the man with the right vision at the right time.

Thomas Newcomen's steam pumping engine had first been demonstrated in 1712 and, in very short order, the steam engine was recognised as the future driver of Britain's industrial development. Steam engines needed coal, and the road system was not yet sufficiently developed to meet the challenge of moving ever larger quantities of coal from the collieries to where the engines were.

First to explore the idea of using a canal to link major coal producers was Francis Egerton, 3rd Duke of Bridgewater who owned extensive mines and had a growing number of end-users in Manchester and Liverpool. He saw huge commerical benefits in building a canal linking his collieries at Worsley in south Lancashire with the north-west's two most rapidly growing conurbations, Manchester and Liverpool.

The Bridgewater Canal was Britain's first 'pure' canal, cutting a completely new path through the countryside. Earlier canals had either been 'canalised' sections of rivers to make them more easily navigable, or linked rivers via short canalised sections.

The Bridgewater Canal was designed as a 'Contour Canal' – following a level course along its route, obviating the need for locks. Contour canals were much quicker to navigate than later canals with locks, but they often had to follow rather circuitous routes. They also presented unique challenges. Those which Brindley faced with the Bridgewater Canal would lead him to revisit and re-imagine aspects of Roman engineering.

At the time the Duke was planning his waterway, Brindley was already surveying routes for the Trent & Mersey Navigation which was being promoted by the Earl of Gower, the Duke's brother-in-law, and Egerton persuaded him to take on the challenge of the new canal. He met the Duke and his advisers at Worsley Old Hall – the first of many such meetings – and was invited to take on the job. For the duration of the surveying and planning stages, he regularly stayed at the Hall.

There were plenty of doubters who questioned whether or not such a canal could ever be built – especially as it had to incorporate what would be Britain's first canal aqueduct if it was going cross the River Irwell at Barton-upon-Irwell.

[above] Worsley Packet House on the Duke of Bridgewater's canal. In 1759 planning meetings for the canal were held nearby in the Duke's Worsley Old Hall. The spur off the canal to the right of the photograph gave access to Worsley Delph and 47 miles of underground canals, on four different levels with water-powered inclined planes, allowing the Duke to mine around a million tons of coal a year and deliver it along the waterway to the mills of Manchester and elsewhere. The orange colour of the water comes from iron oxide leaching from the coal mines.

[right] This view of Brindley's Barton Aqueduct, photographer unknown, was used in 1894 by local artist Albert Dunington (1860–1941) as the basis for his oil painting of the bridge which is now part of the Permanent Collection in Salford Art Gallery.

The Romans had engineered aqueducts to carry water great distances, but the idea of using one as part of a navigable waterway was original. When it was opened, large crowds are said to have gathered on the banks of the Irwell to stand in wonder as the first boats sailed across. Brindley is said to have shown the Duke that such an aqueduct would be entirely feasible by carving one out of cheese and demonstrating that it could hold water without leakages.

Initially there were problems with the sheer weight of water the bridge was carrying, requiring the masonry to be strengthened, but once rectified, it operate successfully from 1761 until replaced by Edward Leader Williams' Barton Swing Aqueduct in 1893. Parts of Brindley's original bridge were built into Williams' replacement.

The canal between Worsley and Stretford opened to traffic on 17 July 1761, and four years later it was extended to Castlefield Wharf in central Manchester. Its opening initiated a period of great success for the Duke, and growing wealth, but it would take another 11 years before his dream of extending the canal to the Mersey at Runcorn could be realised. Brindley never saw it completed, having died in 1772.

To accommodate the change of levels on the Runcorn section, a flight of ten locks was required. Locks had long fascinated Brindley – the idea of taking a boat up one side of a hill and down the other simply by moving water would be the key to the development of an industrial transport infrastructure long before the railway era.

Brindley effectively defined the size and capacity of locks – and thus the size of canal boats – a single lock to take a narrowboat would measure 72ft. 7ins. (22.12m) long by 7ft. 6ins. (2.29 m) wide while a double-width lock would be 15ft. wide (4.6m) and able to accommodate two narrowboats side by side or a single wide boat.

The single lock used less than half the water required by a double-width lock, and in parts of the country where feeder streams were in relatively short supply, that was a significant benefit, albeit limiting boat size substantially. To accommodate heavier traffic, wider stretches were engineered at either end of a flight of locks – a space where boats could wait their turn.

Brindley's vision for the canal system which he proposed building was both innovative and challenging. It involved what he referred to as his 'Grand Cross' which he believed would eventually link the Trent, the Mersey, the Severn and the Thames, but his backers – who included Josiah Wedgwood – were more interested in linking the growing pottery towns and villages to the the potential export opportunities in the 'New World' which would open up once they could get their goods to the port of Liverpool, which already had established trading links with America and Canada.

[above left] Brindley's aqueduct was demolished in 1893 when the Manchester Ship Canal was being built, to make way for Sir Edward Leader Williams' Barton Swing Aqueduct, seen here under construction in a photograph taken for the journal *The Engineer*.

[above] Leader Williams' swing aqueduct opened in 1894 and featured in a number of postcards published in the Edwardian era. This view was originally published as a postcard in 1902. The bridge is still in regular use today.

[below] Brindley, Boulton, Watt and others were among the 'Pioneers of the Industrial Revolution' featured on a series of Royal Mail stamps in 2009.

The canal network would eventually extend to over 4,000 miles becoming the motorways of the day. In a 13-year period Brindley was involved in the construction of 360 miles of waterways.

A key part of Brindley's vision for a nationwide canal network was his Oxford Canal, which he surveyed and designed but never saw through to completion, having died only three years into the project. From the outset, the project was under-funded, but the ever-inventive Brindley recognised early on that many of the crossing points would only ever encounter light use. Thus, instead of the considerable expense of stone-arch bridges, he designed a series of 20 lightweight wooden 'lifting bridges' which could be raised and lowered quickly by hand when access across the canal was needed.

By the early 19th century, it had become possible to reach most of the industrial centres of England by canal boat, and thus transport industrial output nationwide. To achieve that, engineers had embraced the very latest in late-18th century technology – the steam pumping engine.

The primitive steam engine now usually known as the 'Smethwick Engine', restored and preserved in Birmingham's Thinktank Museum, is the closest we will get to seeing how the steam revolution was born. Newcomen engines were already in widespread use 60 years before Watt built his engine, but it was the Smethwick Engine which really demonstrated that steam power could be economically viable. It was built in 1779, designed by James Watt to pump water into Brindley's Birmingham Canal, and was one of a pair of engines specifically installed for the purpose.

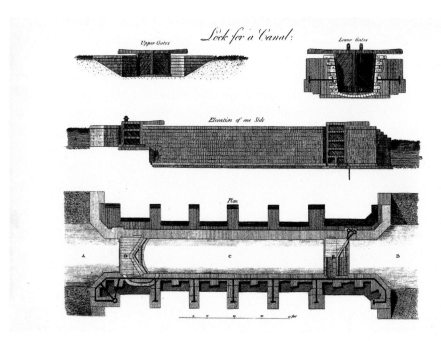

Lock for a Canal:

Upper Gates   Lower Gates

Elevation of one Side

Plan

[left] Brindley's design for the size and construction of a canal lock became the standard template for subsequent canal builders, thus dictating the size of the narrowboats.

[below] Thomas Telford and William Jessop adopted a 'water-saving' design for their canal at Ellesmere Port by having two lock systems side by side. One could accommodate two narrowboats side by side, while the left-hand system was built to Brindley's original narrow design, wide enough for just a single boat, thus using much less water for each operation of the lock.

Whilst Brindley had recognised early in 1768 that the nature of the landscape through which his canal would pass would require a number of locks – and therefore a feeder supply of water – he had under-estimated the demand which heavy usage of the canal would place on his infrastructure. He predicted the future when he noted that

> 'The canal through Smethwick will have to go over the hill using locks and steam pumping engines.'

He knew that the most direct way to build such a canal would have been by cutting a tunnel but the geology of the area – and engineering know-how of the time – ruled that out, requiring the construction of a flight of six locks up one side of the hill and six back down the other side. But not only had he under-estimated the amount of water his canal would need, it also became apparent within a few years that the Newcomen engines which had been installed were inadequate, and that the challenge required a more efficient engine than any yet manufactured.

Whilst he had installed engines to pump water in mines and on other waterways, the Smethwick canal was going to need pumps which could supply much more, and it would not be until several years after his death that action was taken to design and build what would briefly become the world's most powerful steam engine.

To supply the locks with water, Brindley had correctly envisaged pumping water from the lower locks up to a large reservoir at the summit, but the reservoir turned out to be nothing like big enough. Demand rapidly exceeded the capacity of the locks, and by 1790 a new cutting had been dug under the supervision of John Smeaton, taking the top three locks on either side of the summit out of the equation, and adding a much bigger supply reservoir.

The engine then had to raise water just half the original height – 20ft (6m) instead of 40ft (12m) – and at the same time it was fitted with a larger pump to further increase capacity.

Smeaton's canal later became known as the 'Old Main Line' of the Birmingham Canal Navigation, and a few years later, the route was radically redesigned by Thomas Telford, creating the 'New Main Line' which is still in use today. A short section at the bottom of

[opposite page top] Brindley's narrow Spon Lane Lock on his Birmingham Canal opened in 1779. To the right is Thomas Telford's New Main Line route, which opened in 1827, halving the Smethwick lock flight.

[middle & bottom] Brindley built wooden lift bridges as an inexpensive way for lightly used rights of way to cross the South Oxford Canal. Although they are now operated either by hydraulics or windlasses, recent replacements are still true to Brindley's original design. This is Drinkwater's Lift Bridge, Bridge No.231, *middle*, and Shipton-on-Cherwell Bridge, Bridge No.219, *bottom*.

Brindley's original flight, the Spon Lane Locks, can still be seen, and while traces of its original route are still observable in the landscape, the majority of the canal now runs at Telford's level, many feet below Brindley's.

Josiah Wedgwood, one of the canal's original backers, had recognised from the outset of the proposal that the potential benefits of a canal network in shipping ceramics from his factory were enormous. Matthew Boulton, another of the project's sponsors, who planned to use it to move materials into, and finished goods out of, his Smethwick 'manufactory', supported the idea of installing the first of a pair of large pumping engines to supply the shortfall, and suggested there was nobody better to design and build them than his new business partner, James Watt.

The canal itself, completed just weeks before Brindley's death in 1772, ran 24 miles from Birmingham to the Staffordshire & Worcestershire Canal, through Wolverhampton and Smethwick where Boulton's Soho Manufactory had been established in 1766. In 1775, just three years after the canal opened to commercial traffic, the historically important partnership of Boulton & Watt was formed.

[right] Loading salt on to barges at a Northwich salt works, c.1905. The shipping of salt was one of the main cargoes on the Trent and Mersey Canal, which was completed just after Brindley's death. Until 1871, the salt was unloaded at Anderton and sent 50 feet down chutes to waiting boats on the Weaver Navigation for shipping further afield.

[below] After 1871, the fully-laden boats were lowered vertically from the canal to the Weaver on Edward Leader Williams' innovative Anderton Boat Lift. Williams would later serve as chief designer and engineer on the building of the Manchester Ship Canal.

[below right] 'Facsimile of Brindley's Hand-writing', from his domestic accounts, one of the illustrations in Volume One of Samuel Smiles' *Lives of the Engineers*.

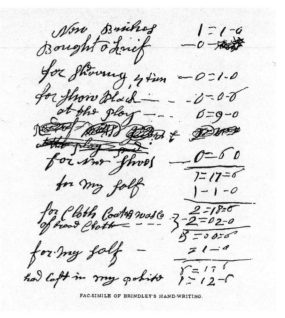

FAC-SIMILE OF BRINDLEY'S HAND-WRITING.

When Watt completed the previously-mentioned Smethwick Engine in 1779, it instantly became the most efficient steam engine yet built. It went into service that same year and marked the beginning of the partnership between the ingenuity of James Watt and the entrepreneurship of Matthew Boulton which revolutionised industry and is explored later in this book.

Perhaps Brindley's most remarkable achievement – with his health already fading and with critics immediately prophesying that this would bring about his downfall – was the cutting of the Harecastle Tunnel under Tunstall Hill near Kidsgrove on what we know today as the Trent and Mersey Canal, a key component of his 'Grand Trunk Canal' vision. At more than a mile and a half long – for a time the longest tunnel on the canal network – this was a huge undertaking, stretching engineering know-how to its limits It was excavated by teams of navvies working out from the bottom of numerous vertical shafts across the hillside, their working environment prone to regular flooding until steam pumps from Boulton and Watt were installed. Construction took seven years, but Brindley only saw the first two years, dying in 1772, his role as Chief Engineer being taken on by Hugh Henshall, (1734–1816) who had worked with both Brindley and Smeaton. For nearly 140 years, lying on the roofs of their barges with their feet on the walls, boatmen 'legged' their boats through the tunnel in total darkness.

When Brindley's health began to fail, it was his friend Erasmus Darwin – Charles Darwin's grandfather – who diagnosed diabetes which, together with pneumonia, brought about his premature death at the age of just 56.

[top left] The entrance to Brindley's 1777-built 2,880 yard (2,630 metres) long Harecastle Tunnel on the Trent and Mersey canal. Telford's much wider tunnel opened in 1827. Brindley's tunnel closed to traffic in 1914 after a partial roof collapse.

[top right] Brindley and Robert Whitworth surveyed the route for the Leeds and Liverpool Canal. The first Chief Engineer was John Smeaton's pupil John Longbottom who resigned in 1775 and was succeeded by another of Smeaton's pupils, Robert Whitworth who had also worked with Brindley.

[above] A late 1770s aquatint of the Trent and Mersey Canal near Burslem. A spur to Burslem would be created in the early 19th century.

Superimposed on to a late 19th century chromo-
lithographed view by the Photochrom Company of
Zurinch of the fourth Eddystone Lighthouse and the
stump of Smeaton's tower, is an 18th century cross-
section view of his revolutionary design. Smeaton's
1759 lighthouse, which stood solid against the storms
for 123 years, was 72ft (22m) high, less than half
the height of its replacement at 161ft (49m). The
fourth (1882) lighthouse, which still stands today, was
designed by James Douglass, employing refinements
of Smeaton's pioneering construction techniques
developed by Robert Stevenson.

JOHN SMEATON, THE FIRST CIVIL ENGINEER

[above] An engraving of John Smeaton, with the Eddystone Lighthouse behind.

[below left] Smeaton's Tay Bridge in Perth was opened in 1771 and widened in 1869. It is still used today and is Scotland's only 'Grade A' listed bridge.

[below] The lighthouse featured in the background, behind Britannia, on the 1860 penny coin and on most mintings of the old penny until 1970.

Travelling by coach from Glasgow to Stirling in 1724, Daniel Defoe, the English writer and journalist was struck by how close Scotland's two great rivers, the Forth and the Clyde, came to each other at their closest point. Building a canal between them should, he suggested, be a simple undertaking.

In his book *A Tour Thro' the Whole ISLAND of GREAT BRITAIN* he referred to the trading benefits of some of the canals which had been built in France, and concluded 'I leave it to time, and the fate of Scotland, which, I am perswaded, will one time or other bring it to pass.' Defoe was undoubtedly a great writer, but he was certainly no engineer. He believed his canal could:

> 'be done without any considerable obstruction; so that there would not need above four sluices in the whole way, and those only to head a bason, or receptacle, to contain a flash, or flush of water to push on the vessels this way or that, as occasion requir'd, not to stop them to raise or let fall, as in the case of locks in other rivers.'

Forty years later, a much more viable plan to link the Forth and Clyde was proposed by Leeds-born engineer John Smeaton (1724–1792), often referred to as the 'Father of Civil Engineering' and work started in 1768. However, the canal which Smeaton built – the first in Scotland –

connections with:

- William Jessop
- The Lighthouse Stevensons
- John Rennie
- Thomas Telford
- James Brindley
- Sir William Armstrong

[top] Canal Street in Grangemouth, at the eastern end of the Forth and Clyde Canal, from a postcard c.1905.

[above] The Smeatonian Society of Civil Engineers, the oldest engineering society in the world, was established in 1771. The role of an annually-elected president was introduced in 1841, and holders of the position have included Bryan Donkin, Sir John Rennie, Robert Stephenson, Charles Blacker Vignoles and Joseph Bazelgette. The third series of Smeaton Medals, awarded for outstanding achievements in engineering in hostile environments, was struck in 2021 to mark the 250th anniversary of the society's foundation, and features a motif of the Eddystone Lighthouse. The medal is now presented annually. (*courtesy of the Smeatonian Society of Civil Engineers*).

would not be eight miles long as Defoe had confidently suggested, but 35. And rather than having no locks, it required 39 of them.

Over a long and eminent career which embraced physics, chemistry and both mechanical and civil engineering, Smeaton built canals, lighthouses, bridges, harbours and watermills amongst other things. To him fell the challenge of building the canal to link the two rivers

Work started on the project with the passing of an Act of Parliament in 1768, and the project's main backer, Sir Lawrence Dundas, wrote a few years later that

'the execution of this canal proceeded with such rapidity, under the direction of Mr. Smeaton, that in two years and three quarters from the date of the first Act, one half of the work was finished; when, in consequence of some misunderstanding between him and the proprietors, he declined any further connection with the work.'

After Smeaton left the project, it would take 22 more years to complete and would not open until 1790. By then, the role of engineer on the project had passed to Robert Whitworth, a protegé of Brindley, who saw it through to completion. Smeaton had moved on to other, greater, projects.

With the increasing size of vessels, it never became the busy shipping lane its backers had envisaged, and long before traffic on it was brought to an end by the construction of central Scotland's first motorways in the 1960s, it had become mainly used by coastal craft seeking a short cut.

Today, thanks to the building of the spectacular Falkirk Wheel – which replaced a flight of 11 locks designed by Hugh Baird and opened in 1822 – the canal is once again linked to the Union Canal giving access to Edinburgh. Now predominantly used by leisure traffic, it has become the focal point of a resurgent tourist industry.

When the route was first surveyed and budgeted, Smeaton believed it would cost £147,337 – a very precise sum. Despite all the delays, and changes to engineers and contractors, the final cost was just a little more than £2,000 over budget.

The Forth & Clyde was not Smeaton's first foray into canal-building. He had earlier been involved with Brindley in surveying the route for the Duke of Bridgewater's Canal. Indeed, the two men were friends, and held each other in great esteem, seeking out each other's counsel, and recommending each other for several projects. On Brindley's death in 1772, Smeaton actually took over responsibility for seeing a number of his projects through to completion.

Smeaton was born near Leeds in 1724, and from an early age demonstrated a fascination with all things both mechanical and architectural, despite his father's wish that he should have studied law. Engineering at that time was not yet accepted as a profession, there were no academic courses devoted to the study of the subject, and the only recognised engineers were found within the military. According to Samuel Smiles in his *Lives of the Engineers*, Smeaton's father found it hard to understand why his son would eschew the legal profession for the life of a 'mechanical workman'.

By 1750 and living in London, young Smeaton was mixing in academic circles, attending meetings of the Royal Society and developing a career as a maker of mathematical and scientific instruments. Over the following few years, he delivered several papers to the Royal Society describing improvements to astronomical and maritime instruments, improvements to Thomas

Savery's steam pump, and a geared crane with a pulley system which allowed one man to raise or lower a one ton weight.

Smeaton believed that the subject of engineering was more widely understood and practised in Europe than it was in Britain and, having learned to speak French, he spent several months in 1754 and 1755 visiting Belgium, Holland and France to learn the latest techniques being used in the construction of docks and canals.

But his interests and experiments went far beyond canals, lighthouses and docks. Six years after having been elected a Fellow of the Royal Society at the age of 29, he was awarded the Society's prestigious Copley Gold Medal in 1759 for his thesis *An Experimental Inquiry concerning the Natural Powers of Water and Wind to turn Mills and other Machines depending on a Circular Motion.*

Out of that research grew what became known as the 'Smeaton Coefficient' which would, nearly a century and a half later, play a crucial part in the Wright Brothers' calculations in their preparations for the world's first powered flight.

After a series of tests on optimum configurations, he built the first asymmetrical smock mill in Britain, with five sweeps rather than four. It was in the village of Spital Tongues in Northumberland. While several such mills were built in Yorkshire and Lincolnshire in the 19th century, they proved quite challenging to maintain.

He has often been described as the 'first civil engineer' – a term he originated to distinguish himself from the usual term of 'military engineer' – but such a description of him has proved to be too narrow, as many of his mechanical devices played significant roles in furthering industrial development. Of particular merit were his water pressure engines which used hydraulic power to raise coal from mines, and predated Sir William Armstrong's hydraulic systems by almost a century.

He also designed several important bridges – at Perth, Aberdeen, Coldstream and Hexham – and Smeaton's Viaduct in Nottinghamshire, several other canals including the Calder and Hebble Navigation between 1758 and 1770, the Ripon Canal at around the same time and the Birmingham and Fazeley Canal in the 1780s as well as improvements to harbours from Charlestown in Cornwall to Peterhead and Banff in Scotland.

To most people, however, Smeaton's name is most easily associated with 'Smeaton's Tower', the former Eddystone Lighthouse which had now stood on Plymouth Hoe for more than 130 years – the same length of time it stood on the Eddystone rocks.

Smeaton is rightly celebrated as the father of the modern lighthouse, the Eddystone tower – the third to stand there – quickly being recognised as an exemplar for future lighthouses.

[above left] The great engineers around whose achievements this book is constructed would have marvelled at the efficiency of the Falkirk Wheel. Much of the design and planning of their canals revolved around sourcing water to keep them functioning. So perfectly engineered is the structure, that rotating the wheel – which weighs eighteen hundred tons fully laden – uses no more electricity than six electric kettles as it raises and lowers boats between the Forth & Clyde and Grand Union canals with almost no loss of water.

[above] Smeaton's 5-sweep Chimney Windmill, built in the Northumberland village of Spital Tongues in 1782, replaced an earlier mill from before 1650.

The first had been swept away in a ferocious storm, the second burned down when years of spilled candle wax caught fire.

Both the earlier lighthouses to be built on the Eddystone rocks had been largely built of wood, but Smeaton broke with tradition and planned to build his tower out of granite – quite a challenge on a rocky outcrop 13 miles south-west of Plymouth with the only access being by small 10-ton sailing vessel usually simply identified as the 'Eddystone Boat', but later referred to by Samuel Smiles as the *Neptune*. If the wind was insufficient, the crew had to row the 14 miles to the rocks.

Smeaton knew its construction was going to be physically demanding, so he recruited teams of hardy Cornish tin miners, and to save them from being press-ganged into the navy – still a common practice at the time – they were issued with metal tokens by Trinity House which exempted them from naval service.

The elegant tapered shape of Smeaton's design for the Eddystone Lighthouse was, he said, evolved after studying the shape of an oak tree, and its resilience to storm damage. The circular shape of the tower, he argued, presented the least resistance to the wind, and thus reduced the impact storms might have on the structure.

Some of the larger blocks of stone weighed more than five tons, with the attendant logistical problems in the days before hydraulic cranes and reliable transport, of moving them from the quarry to the masons' yard where they were cut and dressed. The masonry yard was established at Millbay under the direction of Josias Jessop who continued to work for Smeaton for the rest of is life, while his son William became Smeaton's protégé and would go on to do important work with John Rennie, Thomas Telford and several other great engineers.

Josias Jessop was a shipwright and 'Quartermaster' at Plymouth Docks and while his primary role there was to oversee the repair of wooden ships, he was also responsible for maintaining John Rudyerd's wooden lighthouse, the second tower to be built on the Eddystone Rock. He fulfilled that role for nearly 20 years until Rudyerd's tower burned down.

At Millbay the huge blocks of stone were cut and shaped by an army of stonemasons and temporarily assembled to ensure everything fitted together. From the yard they were transported to the quayside to be rowed or sailed to the construction site and then hoisted up on to the rock.

Work on the rock had to take account of the tides, giving the workmen six hours at most, the much of the first season was taken up with preparing the site. Whereas the foundations of the earlier lighthouses had been simply been cemented on to the rock, Smeaton had his workforce spend months cutting dovetails into the rock itself, into which the lowest courses of the tower would be fitted. Thus tower and rock effectively became part of the same structure, inextricably keyed together.

On calm days, work even continued into the hours of darkness with large torches being used to illuminate the site. But for several months of the year, the sea was just too rough for any work to be done on the rock, but work continued at Millbay.

Cutting the dovetails on the rock proved to be a more arduous task than had been anticipated – a combination of the hardness of the rock and the variability of the weather limiting the workmen to no more than two hours per day on the site. Smeaton, however, was determined that the foundations should be completely prepared

A late 19th century chromo-lithograph of Ramsgate Harbour. The success the port enjoyed in the 19th century was directly as a result of Smeaton's advice and recommendations published in his *Historical Report on Ramsgate Harbour* which he prepared for the Trustees and published in 1791 (*see inset*). He was involved in the port's development for 16 years from the mid-1770s until his death in 1792. Key to the port's success was what he described as the 'Advanced Pier' which so successfully protected the inner harbour from storms that he noted in the introduction to his report that 'in *January*, and part of *February* 1790, there were in it no less than 160 Ships and Vessels at one time, that came thither for refuge in distress, and to save the wear and tear of their tackle and furniture, all of which must otherwise have crouded the *Downs*,–Almost an equal number, for the same reason, came into the Harbour, and were in it at one time, during the tempestuous weather of last *January*, amongst which were *four West Indiamen*, from 350 to 500 tons.'

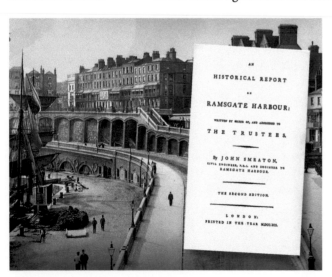

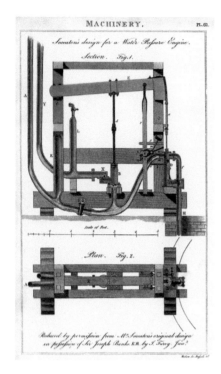

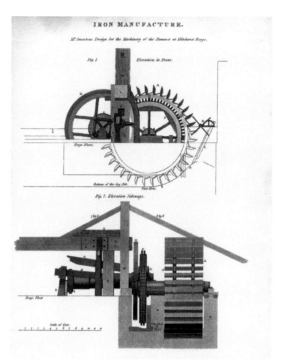

[left] The water-powered hammer which Smeaton designed and installed at Kilnhurst Forge near Rotherham in 1765. In the mid 18th century the village of Kilnhurst had a watermill, two potteries and an ironworks, and there had been a forge since before 1720.

[far left] Published posthumously from the collection of his patron, the botanist and scientist Sir Joseph Banks (1743–1820), another of Smeaton's designs for a water pressure engine. Banks, President of the Royal Society and one of Smeaton's close friends, acquired his archives after the engineer's death.

before the end of 1756 – something which was not achieved until the end of November, four months after work had started.

The first block of stone, weighing around two and a half tons, was ferried out to the rocks on 12 June 1757 and dropped into its pre-prepared position on the sloping, stepped site. Three more stones were laid the following day completing the first course. Thirteen more stones were added before the end of the month, completing the second course.

Samuel Smiles gave a dramatic description of the challenges faced by Smeaton and his workmen – not just on the rocks, but during the often-perilous journeys to and from Plymouth during storms.

As the tower grew taller, working conditions for the builders got a little easier as far as protection from the waves was concerned, but by the 46th course of stone, they were having

One of Smeaton's several designs for water pressure engines which he predicted would have widespread use across the developing industrial landscape.

to raise more than 60 separate blocks up to the working level and then lock them all together. It was not just the intricate pattern of interlocking stones which marked Smeaton's lighthouse as the start of a new era. He also pioneered the use of hydraulic lime mortar – widely used by the ancient Egyptians and Romans – to give the tower its unique strength. Conventional lime mortar sets as a result of carbonation – drawing carbon dioxide from the atmosphere – but thousands of years ago the Egyptians, and later the Romans, found that hydraulic lime mortar was cured by drawing moisture from its surroundings, enabling it to set underwater and thus making it ideal for the lighthouse project.

The secret was using a slaked lime which had a high clay content and, in the case of the mixture Smeaton chose to use, also had the addition of volcanic ash which promoted quick initial setting and high enough strength

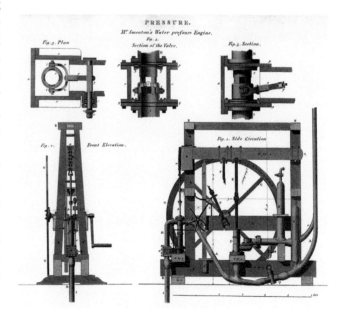

[above] The Kirkintilloch-built SS *May Queen* sailed on Smeaton's Forth and Clyde Canal from 1903–1939.

[above right] The Forth and Clyde Canal today, now a popular leisure waterway. Construction started in 1768 and continued until 1790. The 35 mile canal runs from sea locks at Bowling Basin on the Clyde to the River Carron estuary near today's 'Kelpies', where further sea locks gave access to the Firth of Forth at Grangemouth.

[right] Charlestown Harbour was designed and built by John Smeaton for Charles Rashleigh to meet the shipping demands of his copper mining industry. It opened in 1792 allowing ships to load and unload at all stages of the tide. Through a series of artificial lakes, and specially dug channels, water was brought from several miles away in order to maintain a constant level in the dock. Charlestown Harbour predates William Jessop's larger Bristol Floating Harbour by 20 years. Charlestown harbour, the last operational Georgian harbour in Britain, is now a UNESCO World Heritage site.

to bond the stones together without adding unwanted stresses. Smeaton is credited with triggering the revival of hydraulic mortar and with the development of concretes based on it – again, reviving materials which had been in widespread use almost two millennia before him.

The lighthouse was completed by September 1759 when, on 16 October, the 24 huge candles were lit in the lantern for the first time. Some of the candles weighed around two pounds – 0.9 kilograms – while others are said to have weighed as much as five pounds and each lasted a little over half an hour. There was an alarm clock which sounded every 30 minutes alerting the keepers to carry another two dozen candles up to the lantern. Later keepers were delighted when the candles were eventually replaced by an oil-burner in 1810.

It proved to be a major aid to shipping approaching and leaving the expanding port of Plymouth and the naval base at Devonport, and its innovative construction became the template for dozens of other lighthouses in the century which followed.

The later dynasty of lighthouse-builders, the 'Lighthouse Stevensons' whose structures can still be seen around Scotland's coast, owed their success to evolving the basic blueprint established by Smeaton.

By the 1870s, however, the rock on which Smeaton's lighthouse stood was showing signs of fracturing under the relentless onslaught of the sea causing the tower to shake. After 130 years, the decision was made to replace it.

As ships had got a lot larger since Smeaton's tower was built, and thus needed more warning to steer clear of the rocks, the new lighthouse would be twice the height and thus visible from a much greater distance.

When the decision was made to dismantle Smeaton's tower, the mortar holding the lower courses to the rock proved too strong to demolish – a testament to his decision to use hydraulic lime mortar more than a century earlier. One of Smeaton's most enduring contributions to the industrial development of 18th and 19th century Britain was probably his work on improving hydraulic lime mortars, developments which gave his structures their immense strength and would eventually lead others to develop modern cements and concretes. A quarter of a century after Smeaton's death, it would be another Leeds man, brickmaker Joseph Aspdin, who in 1824 patented what is now known as 'Portland Cement', probably the most important building material in the modern world.

While only the stump of Smeaton's Eddystone lighthouse remains *in situ*, several of his other important civil engineering projects survive, albeit modified over the years. The south harbour at Peterhead was built to his design between 1773 and 1775, extended by John Rennie between 1806 and 1809. Further expansion a few years later saw the north harbour built to a design by Thomas Telford.

His innovative design for Charles Rashleigh's Charlestown Harbour near St. Austell in Cornwall, completed in 1792, stands as an exemplar for the concept of what he referred to as a 'floating harbour' where ships could remain afloat behind lock gates while they loaded and unloaded their cargoes at all stages of the tide.

He had proposed the concept of the 'floating harbour' at Bristol as early as 1765 – others would later refer to such developments as 'wet docks' – an idea which would eventually be realised by his protegé William Jessop in the early years of the 19th century, turning Bristol into a major inland port.

Other engineers and inventors profited from his experiments in the decades which followed, a number of them joining the Smeatonian Society of Civil Engineers, named in honour of his achievements.

[top lef] For a few months in 1882 the new lighthouse stood alongside Smeaton's.

[middle] Part of Smeaton's tower was rebuilt in 1877 on Plymouth Hoe as a tribute.

[above] Smeaton's system of interlocking blocks was copied by numerous later lighthouse builders – including Tompkinson & Company for New Brighton's 1830 Perch Rock lighthouse.

[below] The interlocking pattern of granite blocks devised by Smeaton for the lower courses of the Eddystone tower.

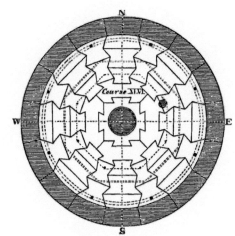

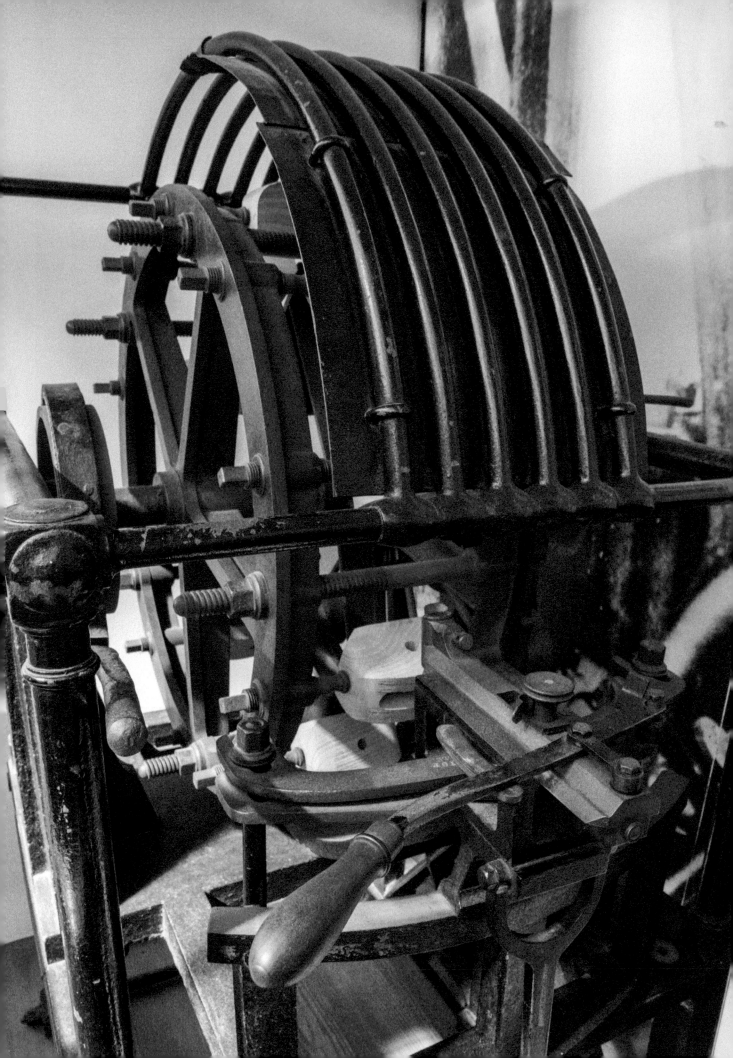

JOSEPH BRAMAH, HENRY MAUDSLAY AND THE DAWN OF THE MACHINE AGE

In popular history, the concept of the production line, or assembly line, is often attributed to Henry Ford, but the introduction of production line methodology predates Ford by more than a century and can be traced back directly to a ground-breaking collaboration between Marc Brunel (1769–1849) – father of Isambard Kingdom Brunel – and Henry Maudslay (1771–1831), now widely considered to be the 'father of machine tools'.

In the closing years of the 18th century, manufacturing was a largely manual undertaking, with the obvious problems of accuracy and consistency. Solving the challenges of achieving consistency drove many innovations in the early years of the 19th century, as did the commercial imperative of increasing output from a largely unskilled workforce. It was that quest for precision and repeatability which led Henry Maudslay towards some of his most important achievements.

Samuel Bentham, the Inspector General of Naval Works at Portsmouth's naval dockyard had a perennial problem to deal with – each of the Royal Navy's sailing vessels used around 1,000 wooden pulley blocks on their rigging, gun carriages and elsewhere on board. A 'First-Rate ship of the line' – at the beginning of the 19th century that was a three-decker with between 70 and 80 guns – would have been equipped with between 1,200 and 1,400 such blocks. These were the origins of the term 'blocks and tackle'.

Because they were handmade, there were inevitable variations in size and quality despite being made by skilled craftsmen. Ropes passing through imperfectly finished blocks created friction, and if the holes were even a little bit too narrow, the wooden blocks – usually made of elm, oak or 'lignum vitae' – could burst into flames. Lignum vitae, also known as 'ironwood', was the preferred choice because of its resistance to wear. Even if they were well made with no rough edges, the more use they got, the faster they wore out.

At the beginning of the 19th century, with the navy expanding to meet the growing challenges of policing Britain's growing empire, the annual demand for replacement pulley

[above] Henry Maudslay, from an 1821 engraving by Parisian artist Henri (Pierre Louis) Grevedon (1776–1860).

[below left] Blocks and tackle on the 1993 replica of James Cook's 1764-built barque, *Endeavour* – now in the Australian National Maritime Museum in Sydney.

[below right] Ropes and pulley blocks used to tether the gun carriages on the 1860-built HMS *Warrior*, the first of the Royal Navy's ironclad ships.

[opposite] Henry Maudslay's Shaping Engine – part of the 'production line' in Portsmouth's Block Mills – shaped ten pulley blocks at a time and was in use from 1804 until 1968.

connections with:

• James Watt
• James Nasmyth
• William Fairbairn
• George Stephenson
• Isambard Kingdom Brunel
• William Armstrong

[above] After the corners have been sawn off, a hole is first drilled in the block of wood in the Boring Machine before being transferred to this machine – the 'Morticer' – in which the hole in routed out and enlarged into the slot through which the rope will eventually pass.

[above right] In the 'Shaping Engine' the blocks – traditionally made of lignum vitae, also known as 'ironwood', or elm or oak – were formed and sanded. Teak blocks were also used in some parts of the world. Lignum vitae is now largely only found in Jamaica and is considered an endangered material.

blocks exceeded 100,000, and at the peak of production, fulfilling the task provided full time employment for more than 100 men.

Marc Brunel – at the leading edge of technological innovation for his day – believed the process could be mechanised, and patented his designs for machines to complete the many stages of manufacture. On 9 March 1801, he was granted a patent – No.2478 *A New and Useful Machine for Cutting One or More Mortices Forming the Sides of and Cutting the Pin-Hole of the Shells of Blocks, and for Turning and Boring the Shivers, and Fitting and Fixing the Coak Therein.*

To develop his ideas into the 45 precision machines, he turned to the 30-year old engineer, Henry Maudslay, whose reputation for engineering precision was already growing. In transforming Brunel's drawings into workable machines, Maudslay introduced extensive modifications, considerably improving them.

In total, there would be 22 different types of machine – saws, mortisers, milling machines and lathes – powered by a 30-horsepower (22.8kW) rotative steam engine, supplied by Fenton, Murray & Wood. Each machine was designed to carry out just one task along the way to producing the finished block, and there were three sets of them, producing the three sizes most ships utilised.

One of Maudslay's first – and perhaps most important – decisions, was to replace the wooden framework for the machines, proposed by Brunel, with iron, giving each machine a stronger, less bulky and more rigid structure. That, in turn, was a key factor in ensuring accuracy in the machines' function and output, and a level of repeatability in the manufacturing process which had hitherto been impossible. The challenge reportedly took nearly six years to achieve, but by 1808 the Block Mills in Portsmouth Dockyard were in full production, peaking at 130,000 blocks that year.

What had hitherto taken a 100 skilled men, now required a team of just ten semi-skilled and unskilled workers, putting those 100 craftsmen out of work – it was the start of production line mass manufacturing in one of the first factories entirely powered by steam.

Henry Maudslay was born in 1771 in a small house in a courtyard behind The Salutation Inn, opposite the main gates to Woolwich Arsenal. His father was a mechanic at the Arsenal – and by the age of 12 Henry was also working there, filling cartridges with gunpowder. By the age of 18, he had been persuaded to go and work with Joseph Bramah, a locksmith who was developing a considerable reputation for the quality of his locks.

Bramah was a prolific inventor, his 18 patents between 1778 and 1814 covered a broad spectrum of inventions, from the development of a float valve for flushing toilets, to a patent remedy for dry rot. In between were patents covering mains water supply systems, improved steam engines and boilers, powered vehicles, and the Bramah lock, considered to be the most secure of the age.

Bramah, who had patented his lock in 1784 (British Patent No.1430) had challenged anyone to try and pick it, with a promise that if they succeeded he would give them 200 guineas. It would be the 1850s before anyone did manage to open it – and then only after more than 40 hours of trial and error over several days. There was a widely-circulated suggestion that the means used to open it were not really in the spirit of the wager.

In the late 1780s, Bramah's locks were expensive and time-consuming to build, and he knew that success would depend on speeding up manufacture, and reducing the price. That it should be Henry Maudslay to whom he turned to help realise those ambitions says much for the growing reputation of the 18 year-old.

Once in Bramah's employment, his very evident skills quickly earned the admiration of both his master and his work colleagues. By studying Bramah's lock, and analysing and rationalising its complexity and the costs of building it, his solution was radical. Samuel Smiles summed up both the challenge and the solution with typical eloquence in his 1863 book *Industrial Biography*.

'Mere handicraft, however skilled, could not secure the requisite precision of workmanship; nor could the parts be turned out in sufficient quantity to meet any large demand. It was therefore requisite to devise machine-tools which should not blunder, nor turn out imperfect work;—machines, in short, which should be in a great measure independent of dexterity of individual workmen, but which should unerringly labour in their prescribed track, and do the work set them, even in the minutest details, after the methods designed by their inventor. In this department Maudslay was eminently successful, and to his laborious ingenuity, as first displayed in Bramah's workshops, and afterwards in his own establishment, we unquestionably owe much of the power and accuracy of our present self-acting machines.'

In the 'Scoring Machine', the grooves on which the rope will sit are cut. After this, the finished shell is planed and sanded before being polished by hand. The machines were installed in 1804 and continued to be used until the Block Mills were closed in 1968. No particular skills were required to operate them as the design of the machines ensured that their action was limited by what were known as preset 'stops' and 'formers'. All the machinery was powered, via line shafting, by two 30 hp (22.4kW) steam engines, the first such manufacturing facility in the world.

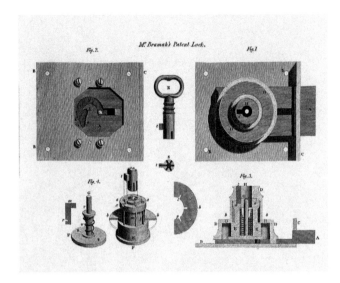

[above] A contemporary illustration of Joseph Bramah's 1784-patented 'unpickable' security lock which was only 'broken' after nearly seventy years

[below] One of a series of plate illustrations of Maudslay's face turning lathes and riveting hammers, from *The Cyclopædia, or, Universal Dictionary of Arts, Sciences, and Literature* by Abraham Rees, published by Longman, Hurst, Rees, Orme & Brown in 1820.

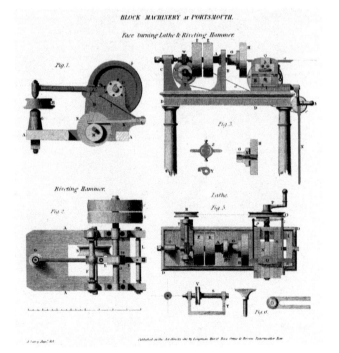

Maudslay did much more than develop the manufacturing process for Bramah's lock – as manager of Bramah's Pimlico Works, he introduced vital modifications which improved the efficiency of the hydraulic press which Bramah had patented in 1795 – and which he referred to as his 'hydrostatic press'. These presses, jacks or rams, would later be used by George Stephenson to raise the tubes of the Britannia Bridge over the Menai Strait and by Isambard Kingdom Brunel and John Scott Russell – with Stephenson's help – to launch the SS *Great Eastern*.

A problem with early hydraulic pumps had been the difficulty of maintaining water pressure, and this was solved by Maudslay's development of a leather 'self-tightening collar' around the piston. Smiles applauded its simplicity and efficacy.

'It was a flash of common-sense genius—beautiful through its very simplicity..... Thus, the former tendency of the water to escape by the side of the piston was by this most simple and elegant self-adjusting contrivance made instrumental to the perfectly efficient action of the machine; and from the moment of its invention the hydraulic press took its place as one of the grandest agents for exercising power in a concentrated and tranquil form.'

According to his narrative in *Industrial Biography*, it was the eminent engineer James Nasmyth who had told him

'Maudslay himself told me, or led me to believe, that it was he who invented the self-tightening collar for the hydraulic press, without which it would never have been a serviceable machine... ...It is the one thing needful that has made it effective in practice. If Maudslay was the inventor of the collar, that one contrivance ought to immortalize him. He used to tell me of it with great gusto, and I have no reason to doubt the correctness of his statement.'

Given his importance in Bramah's successes, it is fair to say that Maudslay believed himself to be worth much more than the reported thirty shillings per week he was being paid. When his request for a pay rise in 1797 was summarily dismissed, he resigned and embarked on establishing his own business, setting up in a small workshop in Wells Street, London.

The contract with the navy to build the block-making machines was an early success for the business, and actually ensured that Maudslay's reputation for innovation and originality of thinking was established

early on – his quest for simplicity, accuracy and repeatability proved to be the key to his commercial success.

Once he had his foundry operational, the Royal Navy contracted him to supply sheets of boiler plate – rolled iron plates used in the manufacture of steam boilers and water tanks. In typical Maudslay fashion, he went further than the contract required. Historically, the plates had been forged using rolling mills, and then sheared to size before being drilled with holes for the rivets – both steps carried out by hand. Thus, the major cost was not in the manufacture of the plate itself, but in its often-imperfect 'finishing'.

According to Smiles – who sadly does not specifiy the size of each sheet – the cost per plate was seven shillings, a considerable sum of money in the early years of the 19th century, especially when considered alongside the 30 shillings a week Bramah had been paying Maudslay just a few years earlier.

After a lecture "On the Self-Acting Tools employed in the Manufacture of Engines, &c." given to the Institution of Civil Engineers on 26 January 1858 by Thomas Spencer Sawyer, the minutes recorded that Maudslay's grandson, Mr. Henry Maudslay, had expressed surprise that

> 'He had not noticed in the Paper any remarks with reference to punching and shearing plates with mathematical accuracy. Nearly forty years ago, his grandfather had invented and worked some ingenious machinery for punching boiler-plates, for making iron water-tanks for the navy... ...When that contract for water-tanks was undertaken, the application of mechanical appliances to such purposes was in its infancy; yet up to the present time, that machinery was almost unrivalled for the precision of its action, and the quantity of work it performed.'

Smiles noted that not only had the seven shillings cost of a finished plate been reduced by more than 90% to just ninepence using Maudslay's machine, but the accuracy of spacing between the holes also made the subsequent work of the riveters a lot less challenging.

Given Maudslay's engineering prowess, and the wide-ranging innovations he pioneered, Smiles does not grant him his rightful place in the pantheon of great engineers, something that, with the benefit of hindsight, we rightly celebrate.

For example, while Marc Brunel patented his tunnelling shield in 1818, he could never have embarked on excavating the Thames Tunnel between Rotherhithe and Wapping in 1825 had Maudslay not developed his ideas into a workable machine. The subsequent development, later in the century of the improved cylindrical shield by James Greathead, made the building

[above left] Maudslay's original 'screw-cutting lathe', photographed in the Lambeth workshops of Maudslay, Sons and Field, for an article in the journal *Engineering* which was published in the issue for 18 January 1901. The company had ceased trading in the previous year. Such was the quality of Maudslay's original equipment that some of the early 1800s tools were still in use.

[above] A machine shop at the Royal Gun Factory at Woolwich, published in *The Illustrated London News*, April 1862, showing the assembly of 'Armstrong Guns', being built under the guidance of Sir William Armstrong (*qv*). Many of the machine tools being used were direct evolutions of those designed more than 50 years earlier by Maudslay who had begun his career as an apprentice in the Arsenal.

of underground railways through difficult geology much less dangerous. The tunnel itself was not a financial success and in 1865 it was acquired by the East London Railway – whose engineer, Sir John Hawkshaw, would later complete Isambard Brunel's Clifton Suspension Bridge.

Amongst Maudslay's many contributions towards the era of precision engineering was his screw-cutting lathe. Before the development of this simple machine on its rigid metal frame in 1800, the cutting of screw threads was a slow and skilled job. It also lacked both precision and repeatability – so much so that each threaded bolt required its own bespoke nut. Maudslay's machine – later refined by Joseph Whitworth who standardised the gauge of threads – meant that nuts could be produced with exactly the same thread count as the bolts they were intended to fit. Maudslay's demonstration piece was a five feet long perfectly-threaded bolt. From those pioneering achievements came the 'quarter Whitworth' bolt still used today.

Despite having patented many of his inventions, Maudslay never sought similar protection for the machine, which would subsequently evolve into the essential precision hole punch for the construction of riveted tanks, boilers, bridges and other iron structures. No illustrations of Maudslay's hole punch have been located so far. Others would profit from his invention, and in turn invent machines which improved the efficiency of the drilling and riveting processes – amongst them William Fairbairn and William Arrol, whose patent riveting machine was crucial in the building of the Forth Bridge.

When Brunel made the decision to build his first transatlantic steamship – the wooden-hulled PS *Great Western* – she was fitted with Maudslay engines as Brunel considered them to be the most reliable. Unlike steamers which had gone before her, sail was intended as her secondary power source, with her steam engines intended to provide the majority of her drive as she crossed the Atlantic.

When launched in Bristol, however, she had no engines, and was sailed round to William Jessop's West India Dock on the Thames to have her engines installed. Maudslay, Sons and Field, would develop an impressive reputation over many years for their marine engines.

[top] Bramah's hydraulic press was patented in 1795 – an early application of the Blaise Pascal's Law of fluid dynamics.

[above] A James Nasmyth (*qv*) steam hammer at Woolwich Arsenal, c.1900. Nasmyth's career had started in 1829 as an apprentice to Maudslay.

[right] The giant facing lathe by Death & Ellwood of Albert Works, Leicester in the workshops at Dinorwig – now the National Slate Museum – can trace its heritage back to Maudslay's original metal-framed bench lathe from 1800.

While Maudslay's growing reputation ensured the success of his Lambeth Foundry and Engine Works, his pioneering machines, installed in the Block Works at Portsmouth in 1804, continued in use for 160 years, until 1968 when the navy's need for blocks had reduced to almost zero. The number had been declining steadily since the introduction of steam instead of sail. The machines were all preserved after they were withdrawn from service and some can be still be seen displayed at Portsmouth Historic Dockyard, while others are in London's Science Museum.

In 2009, Maudslay's many achievements were recognised when he was featured on a Royal Mail 56p stamp, one of a series celebrating the great pioneers of the Industrial Revolution, but it was neither his block-making machines, nor his lathes which were featured. The stamp carried an illustration of one of his compact 'table engines', dating from 1815 – engines which found a multitude of uses within the emerging industrial landscape.

James Nasmyth summed up Maudslay's contribution to the development of the mechanical world by writing

> 'Every member or separate machine in the system of block-machinery is full of Maudslay's presence; and in that machinery, as constructed by him, is to be found the parent of every engineering tool by the aid of which we are now achieving such great things in mechanical construction. To the tools of which Maudslay furnished the prototypes are we mainly indebted for the perfection of our textile machinery, our locomotives, our marine engines, and the various implements of art, or agriculture, and of war.'

[above] The engines for Brunel's transatlantic steamer, SS *Great Western*, were built by Maudslay, Sons, & Field in 1837.

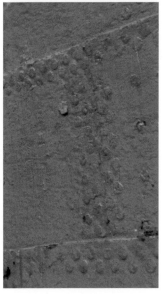

[middle] Hull riveting on Brunel's SS *Great Britain*, launched in 1843. Before the advent of Maudslay's hole punch and its successors, hand punching rivet holes this close together invariably led to inaccuracies and stress fractures whenever the holes on two adjacent plates failed to align correctly.

[far left] Marc Brunel designed the tunnelling shield used to excavate the Thames Tunnel, but once again relied on Maudslay to turn his ideas into a workable reality. Maudslay, Sons and Field also built and installed the steam pumping engines which drained the tunnel works.

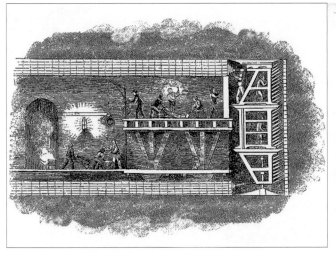

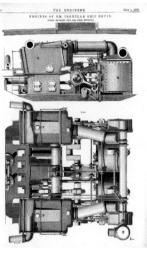

[left] Illustrated in *The Engineer* in 1870 were the compact engines of HMS *Druid*, built by Maudslay, Sons, & Field – mis-spelled in the journal as 'Maudsley'.

WILLIAM JESSOP'S GUIDING HAND

Whilst the names of many of the great engineers of the late 18th and early 19th centuries are immediately recognisable to most people today, the name of William Jessop (1745–1814) is much less familiar. And yet, in his time, he was one of the most highly respected engineers of his generation, and in 1793 was described as 'the first engineer of the Kingdom' at the time of his appointment as Principal Engineer to the Grand Junction Canal Company.

Today, however, his importance seems to have been somewhat overshadowed by history shining a much brighter light on the achievements of those he advised and with whom he collaborated. Perhaps that was the price sometimes paid by those who willingly shared their expertise to help, to improve and to support others, rather than seeking personal fame and fortune.

Even Samuel Smiles afforded him only a few mentions in his *Lives of the Engineers* – an extended footnote covering his childhood under the guardianship of John Smeaton following his father's death, his subsequent ten years in Smeaton's employment, followed in turn by a brief summary of his major engineering achievements in canals and waterways, while his major contribution to the evolution of the railway was given just a single sentence.

[above] A portrait of William Jessop, published in *The Engineer* in 1900.

[below] The development of Bristol as one of the country's great ports can be traced back to the work of John Smeaton and William Jessop. The 1861-built steam tug *Mayflower* – in the centre of this photograph – is the oldest Bristol-built ship afloat and probably the world's oldest surviving tugboat.

[opposite page] Jessop's Floating Harbour in Bristol today – with the Fairbairn steam crane in the distance – seen through the rigging and bowsprit of the replica of Venetian navigator John Cabot's 'Caravel' *Matthew* – the 'little ship of fifty tons burden' on which he sailed from Bristol to Newfoundland in 1497.

connections with:

- James Brindley
- John Smeaton
- John Rennie
- Thomas Telford
- Isambard Kingdom Brunel
- Sir William Armstrong

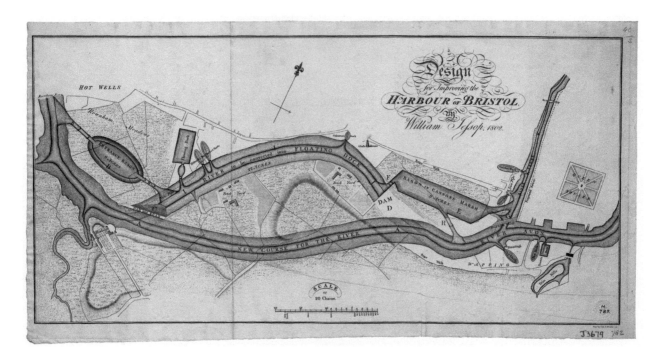

Jessop's 1802 proposal for re-routing the River Avon and creating Bristol's 'Floating Harbour'. A major engineering undertaking which involved dams, cuttings, locks and sluices. It revolutionised the fortunes of the city and much of his design still survives today, albeit extensively adapted and modernised over the past two centuries. Bristol's major role as a commercial port may have been taken over by Avonmouth on the Bristol Channel – due to the vastly increaed size of cargo vessels – but Jessop's harbour is now a major leisure facilitiy. *Courtesy of Bristol Culture and Creative Industries, M Shed Museum Collection.*

'In 1789, Mr. William Jessop constructed a railway at Loughborough, in Leicestershire, and there introduced the cast-iron edge rail, with flanches cast upon the tire of the wagon-wheels to keep them on the track, instead of having the margin or flinch cast upon the rail itself; and this plan was shortly after adopted in other place.'

That seems inadequate for a man who worked with John Smeaton, John Rennie and Thomas Telford, and whose ideas influenced many of his contemporaries.

So who was William Jessop, and how did he go from being 'among the most eminent engineers of his day' as Smiles' footnote described him, to being largely overlooked?

He was born in Devon in 1745, the elder son of Josias Jessop (1710–1760) and his 19-year old wife Elizabeth Foot. John Smeaton engaged Josias as his foreman and overseer on the building of his Eddystone lighthouse, and the men developed a professional bond and a close friendship.

The young Jessop was just 14 when that great project was completed in 1759, the year in which Smeaton took him on as an apprentice draughtsman, and Jessop moved to Austhorpe Lodge near Leeds, the Yorkshire house where Smeaton had been born and where he lived his entire life.

Less than two years later Josias Jessop died and out of respect for his valued colleague and friend, Smeaton continued to support the family in a number of ways throughout the remainder of William's apprenticeship until its completion in 1767.

For the following five years William worked as his assistant and gained much practical experience in marine engineering while Smeaton executed several harbour and waterway projects. Those projects included work on the Forth and Clyde Canal and the Calder Navigation. In 1773, by then working on his own behalf, he accompanied Smeaton to Ireland to advise on solving problems with the Grand Canal which was being built to connect Dublin with the River Shannon.

That same year, Jessop was elected a member of the Society of Civil Engineers which had been established by Smeaton two years earlier. He later become the Society's secretary, a post he held until 1792 – the year Smeaton died – but, perhaps surprisingly, the society ceased to function before the end of that year. When it was reconstituted in 1793, the minutes record that

Broad Quay as it was c.1875 looking towards St. Augustine's Reach on the River Frome – the part of the floating harbour closest to the city centre. In the distance is the tower of St. Stephen's Church. As ships got larger, these docks fell out of use and the reach was in-filled and the river canalised in the early 20th century. Entrance to the reach and its quays was opposite today's M Shed Museum.

[below left] The Remains of Jessop's old North Entrance Lock looking in towards the Floating Harbour, seen here at very low tide with the Plimsoll Swing Bridge beyond.

[below right] The PS *Medway Queen* in 2013, with her new Heritage Lottery Funded hull, after being floated out of David Abels' Albion Dry Dock in Bristol's Floating Harbour, near the Great Western Dry Dock where Brunel's SS *Great Britain* is now preserved.

'The first meeting of this new institution The Society of Civil Engineers was held on the 15th of April 1793 by Mr. Jessop, Mr. Mylne, Mr. Rennie and Mr. Whitworth'

Its aims were defined as 'promoting and communicating every branch of knowledge useful and necessary to the various and important branches of public and private works in civil engineering', and shortly thereafter the society changed its name to the Smeatonian Society of Civil Engineers in honour of its founder. Today it is the oldest society of engineers in the world, its membership embracing a broad church of engineering specialisms.

Smeaton had introduced the title of 'civil engineer' into everyday language, a description of himself which Jessop readily adopted and used widely to promote his own abilities and his ideas for the development of a more structured approach to the planning and implementation of improvements to the country's transport infrastructure.

Jessop was one of several civil engineers in the 1780s arguing for the instigation of a national geographical survey and the placement of 'benchmark' stones to aid future surveys. Surveying and mapping had hitherto been the province of military engineers.

He understood that, especially when seeking to establish reliable water feeder systems for his canals, a knowledge both of the contours through which the water flowed, and the riverbed materials through and over which it flowed, were key areas for research and analysis which needed to be better understood. Some of his proposals about the benefits of establishing holding lagoons on canals would be implemented to great and lasting effect by John Rennie on his Caen Hill flight of locks on the Kennet and Avon Canal near Devizes in Wiltshire.

When working on the River Trent Navigation in 1782, he prepared a paper outlining some of his observations – most notably exploring the different effects on both river flow and what we today would call the local eco-structure of widening navigable channels in rivers flowing through different materials. His work in this area led him to see his role as being as much about the management of the river itself as about improving its commercial potential.

Several years later, while working on the Cromford Canal with Benjamin Outram between 1789 and 1794, he used similar techniques to calculate the flow of the fast-moving River Derwent in Derbyshire. The first Ordnance Survey maps appeared in 1801, but Jessop's suggestion about contoured maps, which would benefit all future transport network developments, were not acted upon until the early 1850s.

Jessop was, undoubtedly, a very busy man, and his expertise and innovations much in demand for the many major projects which filled the closing years of the eighteenth century. Some of them, which he might have liked to take on, had to be turned away due to pressures on his time.

The proposers of the Rochdale Canal in Lancashire – which was to run from the Calder and Hebble Navigation at Sowerby Bridge to the Bridgewater Canal in Manchester had first approached John Smeaton in the 1760s to survey a route and design the waterway, but the project foundered. When the idea was revived in 1791, they approached Jessop to survey a number of possible routes and advise on designs but he was fully committed on other projects and turned them away. John Rennie was appointed in his stead, but the project eventually foundered.

When it was revived yet again in 1794, Jessop was available and accepted the commission – considering it a significant challenge as the proposed route had more variations in height than he had ever previously experienced, and he proposed to keep the canal fed with water from reservoirs along the route.

[below] The Rochdale Canal opened in 1804 and was hugely successful, but by the 1860s, it had already lost traffic to the railways. By 1965 it was derelict, and parliamentary powers were granted to completely abandon it.

[bottom] The Rochdale Canal at Hebden Bridge today. Thanks to volunteers and conservationists – and several million pounds of funding – the entire canal has been restored since the early 1990s and has become a popular leisure attraction.

So undulating was the landscape through which it was cut, it required 91 locks along its 31-mile length. However, from the outset, it was a financial success, and many successful industries sprung up along its towpaths. It became the primary transportation route between Lancashire and Yorkshire, and by 1840 was carrying more than 750,000 tons a year – wool, cotton, timber and coal making up the major tonnage.

It survived as a working waterway into the middle of the twentieth century, but an Act of Parliament of 1965 actually authorised its demolition. Luckily that did not happen, and enthusiasts rallied to restore it and stop further deterioration.

Now, after a massive restoration programme in the 1990s, it has become a major leisure facility, with parks, and walks along much of its length, and even cruising holidays for those with the energy to deal with all the locks.

His most enduring projects were concerned with dock and harbour – most notably the creation of Bristol's 'Floating

Harbour' which transformed the city's fortunes by creating a harbour where ships could load and unload at all stages of the tide. The idea had first been proposed by his mentor, John Smeaton in the mid-1760s, but he had never carried it through. A modified scheme was proposed a few years later by a local shipbuilder, William Champion, but it, too, stalled.

Robert Mylne, a fellow-member of the Smeatonian Society had estimated that the project was likely to be too expensive as it would cost in the region of £65,000 – equivalent to around £11.5M today, a figure which would actually seem to represent outstandingly good value for money.

Elements of Champion's proposals were later incorporated into Jessop's 1788 plans, and planning work – which involved re-routing the course of the River Avon – eventually started in 1790 after the whole project had been given Smeaton's enthusiastic endorsement.

Re-routing the river was fraught with difficulties as it is both fast moving and prone to considerable flooding, but Jessop's and Smeaton's calculations accurately predicted the impact that would have on the dams and embankments which would need to be built if the river's peak flow of around 60,000 tons of water per minute was to be effectively managed.

[above left] A postcard scene, c.1905, on the Ellesmere Canal at Trevor, one of the early projects on which Jessop and Telford worked together. Jessop was Consulting Engineer. Horse-drawn pleasure craft offering trips along the canal and across the Pontcysyllte Aqueduct remained popular into the 20th century.

[above] This 1904 postcard of the Chester and Ellesmere Canal reads 'ON THE CANAL, CHESTER This is the Chester and Ellesmere Canal which here is deeply cut into solid rock, and occupies the site of the old Roman Fosse'.

[left] Thomas Rowlandson and Augustus Pugin's 1810 view of Jessop's West India Import and Export Dock, published in Volume III of Rudolph Ackermann's *Microcosm of London*, a lavishly-produced set of books with one hundred and four coloured aquatint plates. This is plate 92. Built between 1800 and 1806, the docks had the capacity to handle 600 typical cargo ships of the day, but within 40 years, as the age of the much larger steamship dawned, that number was reduced significantly.

In 1803 he appointed his son, Josias, as resident engineer on the project, albeit under his direction, thus relieving him to share his time with other projects while leaving his son in Bristol. His salary was £500 per year, a huge sum for a young man of just 22.

William's work on water flow management on the Trent and Derwent Navigation would play a significant role in the development of the Bristol project. He would eventually see the Floating Harbour completed in 1809 after an expenditure of almost ten times the figure which Mylne had declared to be too expensive back in 1786.

Jessop was already a celebrated engineer before the Bristol project. In the 1790s, as Chief Engineer, he had become involved in planning a canal from Chester to Wrexham – the first section of what, by the 1840s, would be known as the Shropshire Union Canal. In that project he was obliged to work with a young engineer and surveyor who was, reportedly 'unknown' to him at the time – Thomas Telford. He was said to have initially been somewhat displeased by the appointment, claiming that Telford was exaggerating the importance of his role within the project management team.

The 52-year old Jessop had been appointed as 'Chief Engineer', while Telford, 36 at the time, was designated as 'General Agent, Surveyor, Engineer, Architect and Overlooker'. However, the two men clearly found a way of working together harmoniously – despite their quite different characters – as Telford would become Jessop's chosen collaborator on several subsequent projects, and their respect for each other grew significantly.

One of the most significant achievements of their partnership was the Pontycysllte Aqueduct which, although usually attributed to Telford, could not have been successfully completed without Jessop's knowledge and experience.

While working with Telford on the Ellesmere Canal, Jessop was also engaged in his largest project in London, – the West India Import and Export Docks on the Isle of Dogs – London's first secure dock complex, surrounded by a high wall to discourage theft.

John Rennie (*qv*) was his Consulting Engineer, and it was he who, in 1801, organised the acquisition and erection of steam engines to pump water out of the excavations as the docks were being built. The three engines all came from Boulton and Watt, and one of them – at just under 28hp (21kW), and fitted with a 36ins cylinder and a wooden beam – was declared surplus to requirements the following year. It was subsequently sold to the Kennet and Avon Canal Company and installed at Rennie's Crofton Pumping Station near Marlborough where it operated successfully for several years.

The West India Docks were 'wet docks' – the same principle as the 'floating harbour' in Bristol – access to and exit from them was through lock gates into the Thames.

The Import Dock to the north covered 30 acres and measured 2,600ft (792m) from east to west by 510ft (155.5m) north to south – an area of 1.3M sq.ft. or 123,000 sq.m. It had a depth of 23ft (7m). At the time it was by far the largest dock ever built. At the peak of construction, around 1,000,000 bricks per week were being laid, causing what today would be called a 'supply chain problem' as the on-site brickworks could not keep up with demand. The slightly smaller Export Dock to the south covered 24 acres and measured 2,600ft by 400ft (122m) and opened in 1806.

Both docks could be accessed either from Limehouse Reach at the west, or Blackwall Reach to the east. Each pair of lock gates gave access to a basin which then branched to either dock, but as most vessels arriving in the Thames approached from the east, the Blackwall Reach basin was considerably the larger.

The largest of the collaborations between the two engineers was the Caledonian Canal which created a direct shipping route from the west of Scotland to the north-east and avoided the often-treacherous waters of the Pentland Firth off Scotland's north coast. The waterway from Corpach at the head of Loch Linnhe to Clachnaharry on the Beauly Firth north of Inverness, required the excavation of linking canals which produced a continuous waterway through Loch Lochy, Loch Oich, and Loch Ness.

In addition to Jessop and Telford, John Rennie had been asked to survey the possible route and submit estimated costings, but the commission went to Telford and Jessop, whose lower cost estimates, submitted separately, had apparently been quite similar.

Jessop's influence has been recorded across a wide spectrum of improvements in the country's transport infrastructure – his suggestions for improving the durability of road surfaces predated John Macadam's work by more than a decade, and his explanation of the relationship between the breadth of carriage wheels and the degradation of road surface which they caused was perceptive, realising that what was accepted practice at the time when repairing roads was inefficient and did not produce a sustainable surface. The stones being used to repair the surface, he suggested in the 1790s, should be 'no larger than walnuts'.

He also noted that when a carriage or waggon was fully laden, the wheels splayed slightly, meaning that the flat rim of the wheel was no longer in level contact with the road surface, speeding up wear and exacerbating the break-up of the surface. His suggestion that the ideal wheels would be wide cylinders was, of course, theoretically correct but functionally impractical.

As the number of vehicles on the roads increased as Britain industrialised, it would be left to others – most famously John Macadam and Thomas Telford in the early 19th century – to develop more durable road surfaces.

Jessop's friendship and close working relationship with Telford must certainly have ensured that his ideas influenced Telford's thinking as he embarked on his massive nationwide road- and bridge-building programme.

Although his name remains surprisingly obscure when compared with the likes of Telford, Brunel and others, the originality of Jessop's thinking, the importance of his researches, and his many innovations were major influences on the achievements of many of those who followed in his footsteps.

The route of the 18-mile Somersetshire Coal Canal was surveyed in 1794 by Jessop and William Smith under the direction of John Rennie. John Sutcliffe was chief engineer.

[opposite top] A chromo-lithograph of the Pontycysllte Aqueduct, produced in Britain by the Photochrom Company of Zurich and dating from c.1897.

[middle] A steamer, heading south from Loch Ness, in the first of the six locks on the Caledonian Canal at Fort Augustus, heading south towards Loch Oich in 1974. Behind is the 1930s swing bridge by Sir William Arrol & Company which replaced Telford's 1816 original. It was refurbished in 2022.

[bottom] In 1805 Jessop surveyed the 14th century breakwater known as 'The Cobb' which protected the harbour at Lyme Regis in Dorset, and developed a programme of repairs and improvements.

THE LIGHTHOUSE STEVENSONS

The looming presence of the rocky outcrop known as Sumburgh Head, with Robert Stevenson's first Shetland lighthouse at the top, dominates the view as planes approach the airport on Shetland's Mainland. Robert Stevenson was the founding member of a dynasty who became so synonymous with lighthouse building that they are usually collectively referred to as the 'Lighthouse Stevensons'.

John Smeaton's pioneering achievements in lighthouse building offered Robert Stevenson firm foundations for his family's emergence as a true dynasty of lighthouse builders, a huge number of whose towers stand today, still doing the job for which they were built, albeit all now automated with not a lighthouse-keeper in sight.

The Stevenson family – which also included Robert's grandson, the successful author Robert Louis Stevenson – spanned three generations of lighthouse builders who, between them, were responsible for more than 150 lighthouses around the coast of Scotland, and many more in Ireland and the British Empire. They were Robert (1772–1850), three of his sons – Alan (1807–1865), David (1815–1886) and Thomas (1818–1887) – and two of David's sons, David Alan (1854–1938) and Charles Alexander (1855–1950).

Robert came into the profession of lighthouse building as a result of being taken on as assistant to his stepfather, Thomas Smith, who had been appointed as the first Chief Engineer to the newly formed Northern Lighthouse Trust (now known as the Northern Lighthouse Board) – the Scottish equivalent of Trinity House – in 1786.

[opposite page] A Fresnel lens, displayed in the Marine Life Centre at Sumburgh Head. The complex structure of the Fresnel lens gives it the same light-transmitting power of a much larger and heavier solid lens.

[below left] A contemporary engraving showing 'The State of the Works in August 1809' as construction of the tower on Robert Stevenson's Bell Rock progressed.

[below] Bell Rock Lighthouse, from Volume Two of Samuel Smiles' *Lives of the Engineers*.

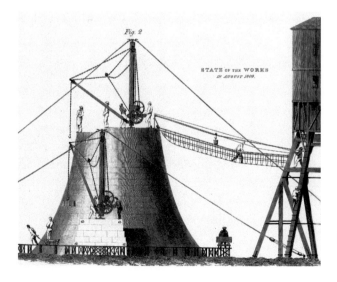

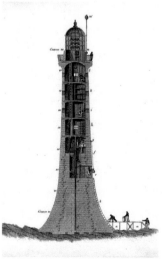

connections with:

- John Smeaton
- John Rennie

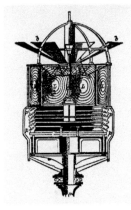

[top] A contemporary portrait of Robert Stevenson, the first of the 'Lighthouse Stevensons'. This portrait is believed to date from the late 1820s.

[above] Stevenson's rotating lens array illustrated in *Encyclopaedia Britannica*

[above right] A contemporary illustration of the tower of the Bell Rock Lighthouse nearing completion, engraved by William Miller after Alexander Carse. Published as Figure xviii in *An Account of the Bell Rock Lighthouse* by Robert Stevenson, published in Edinburgh in 1824 by Archibald Constable and Co. Stevenson's self-balancing tower crane can be seen at the top the tower – a concept which is still used the world over.

Smith was commissioned to build the first modern lighthouses on the Scottish coast, having sought advice and technical expertise from John Smeaton and others. In total, Smith built thirteen lighthouses – ten on his own, and three with his stepson – Cloch in 1797, Inchkeith in 1804 and Start Point in 1806

Robert succeeded him as Chief Engineer in 1808, and by the time he was charged with the design of Sumburgh's lighthouse, completed in 1821, he had already built four others in Scotland, the first having been the remarkable Bell Rock light off the Angus coast which was completed in 1811. Toward Point lighthouse in Cowal had been completed the following year. The Isle of May lighthouse in the Firth of Forth followed in 1816, and Corsewall light near Stranraer in 1817.

Bell Rock was a major achievement – rightly described as 'one of the seven wonders of the industrial world' – and had Robert built nothing else in his lifetime, that achievement would have been sufficient to assure his recognition as one of the greatest engineers of his time.

Nobody had ever contemplated an undertaking like Bell Rock, which involved building a 115ft (35m) high stone lighthouse 11 miles out into the North Sea on a rocky reef which was only visible above the waterline for just a few hours each day. Depending on the tides, there could be just four hours' working time in a 24 hour period, and at high tide, the foundations would be as much as 12ft below the waves. – an inhospitable environment in which to build a lighthouse.

Stevenson first suggested that a lighthouse could be built on the rock in 1799 when he was just 27 years old, and his youth and relative inexperience meant the Northern Lighthouse Board remained unconvinced – that is until the loss of the 64-gun HMS *York* on the rocks in 1804 with the recorded loss of more than 490 crewmen.

Three years earlier, Stevenson had visited the Eddystone Lighthouse and learned much about John Smeaton's innovative construction technique of using interlocking stones. He came away convinced that a similar approach could be used on Bell Rock. Using Smeaton's modification of Roman mortar with its added volcanic ash would also ensure that the setting of the mortar being used on the foundations – which would be inundated within hours of being laid – would actually be aided by being under the salty water of the North Sea.

When a contract to build the lighthouse was agreed, it was John Rennie who was appointed Chief Engineer with Stevenson as his assistant. While Stevenson was in day-to-day charge,

[left] Bell Rock Lighthouse, as depicted in 1819 by one of the 19th century's most celebrated British artists, Joseph Mallord William Turner.

[below left] An Edwardian postcard of Cloch Lighthouse, completed in 1797 by Thomas Smith and Robert Stevenson, and engineer John Clarkson, was the first to bear the name of the Stevenson dynasty. It stands on the south shore of the Clyde west of Gourock. It is one of the three built to cover the approaches to the Clyde estuary, the other two being on Little Cumbrae and Toward Point, at the southern tip of the Cowal peninsula. Smith's and Stevenson's oil lantern at Cloch was first lit on 11 August 1797 and replaced in 1829 with an Argand lamp and silvered mirror reflector. An acetylene lamp was introduced around 1900. The lenses floated on a bath of mercury and were turned by a clockwork motor with falling weights. The lighthouse keepers had to wind the mechanism every few hours.

[below] Cloch Lighthouse today. Like all Scottish lighthouses, it is now automated and unmanned.

several of Rennie's suggestions – such as the dovetailed bonds between the stones – became fundamental aspects of the design, and were pivotal in ensuring the stability of the completed tower. Rennie's dovetails were, themselves, an evolution of the pioneering ideas which Smeaton had evolved for the Eddystone tower.

Work started in 1807, with Stevenson and his workforce living on a small sailing vessel which he named *Smeaton* in honour of his mentor and inspiration. The vessel was also used to ferry the cut stones – Aberdeenshire granite – out to the rocks. After a few months of living on the boat, Stevenson decided a more secure base was needed, and constructed a wooden accommodation tower on stilts adjacent to the lighthouse site, connected to the site by a walkway well above high tide level. Thereafter, the *Smeaton* was only used to ferry the cut stones and supplies.

But Stevenson's innovations did not stop there – as the tower increased in height, the challenges of raising the blocks of stone – some weighing up to a ton – became more pressing, and for that he invented the world's first self-balancing tower crane, initiating a concept which is still key to the construction of tall buildings today. As the tower height increased, the crane was raised with it.

While the lighthouse was an undoubted success, judging by the surviving correspondence between the two men, their relationship soured during construction.

When the lighthouse was turned on for the first time on 1 February 1811, it was the tallest rock lighthouse in the world, and already so famous that excursion boat trips were organised out from the Scottish mainland to see the industrial marvel.

Robert's importance in lighthouse design was immense. Amongst his many achievements was the modification of Fresnel's rotating flashing lights atop lighthouse towers, creating a combined flashing and continuous light, adding to the range of recognisable 'signatures' which could be used to distinguish one lighthouse from another. In some lighthouses his flashing lights even superseded his step-father's idea of using two lights on a tower to confirm the light's location – such as at Pladda on Arran, and although not particularly effective in its early days, was a significant milestone. His system, based on one of Fresnel's prototypes, used a simple clockwork mechanism involing a weight on a chain which turned the lens array as it unwound down into the tower.

Robert was a one-time contributor to *Encyclopaedia Britannia*, and as recently as the 1910 edition, his achievements in improving the Fresnel lens array were recognised with this short citation:

[above] Sumburgh Head lighthouse, Shetland, built in 1821.

[below] Thomas Smith's 1794-built Pentland Skerries lighthouse, *right*, with Robert Stevenson's 1833 replacement, *left*.

'In 1835 Mr. Stevenson, in a report to the Northern Lighthouse Board, proposed to add fixed reflecting prisms below the lenses of Fresnel's revolving light, and he communicated this proposal to M. L. Fresnel, who approved of his suggestion, and assisted in carrying out the design in 1843. This combination added, however, but little to the power of the flash, and produced both a periodically flashing and constantly fixed light; but it must be remembered that the prism for fixed lights was the only kind of reflecting prism then known.'

Robert's modification, approved by Fresnel himself, was first introduced in 1843 to his father-in-law Thomas Smith's 1786 Inchkeith lighthouse, replacing a parabolic reflector which Smith had designed decades earlier. Inchkeith was one of three lighthouses Smith and Stevenson worked on jointly, with Stevenson acting as his father-in-law's assistant engineer. This was ten years after he had completed the Lismore lighthouse in 1833 – his last – and passed the lighthouse-building baton on to his sons.

The Inchkeith lens array was a 'first' in another respect as it used glass ground in cylindrical rings rather than the polygonal segments of Fresnel's original design. This technical advance was pioneered by Cookson & Sons of Newcastle.

Bell Rock was a major undertaking in its day – a challenge on a similar scale to Smeaton's Eddystone light. While the original design for the Bell Rock Tower was Stevenson's, before construction even began the plans had been modified to increase its height, and the engineering prowess of John Rennie had been called in by the Northern Lighthouse Board.

So famous did the Stevenson name become in the world of lighthouses that Rennie's input into the project has been rather down-played in the intervening two hundred years.

One of those who worked on Bell Rock, James Dove, was identified in the project records as the 'Foreman Smith'. His brother David was also one of Stevenson's team. James also worked on the Sumburgh light, and later established an architectural engineering business – James Dove & Company of Greenside in Edinburgh – and more than 80 years after the lighthouse was completed, that company would be involved in a major upgrade to it in 1906. Dove & Co had also been involved in the engineering of David and Charles Stevenson's Fair Isle (North) lighthouse in 1892.

The family's third last new lighthouse was built on Shetland's Mainland on the rim of an ancient volcanic crater at Eshaness. It was completed in 1929 by Robert's grandson David Alan Stevenson. David's last tower was at Torness, opened in 1937.

While nothing like as demanding as the construction of Bell Rock, building the lighthouse on Sumburgh Head presented unique challenges because of its exposed position at the top of the 328ft (100m) high rock. It serves as an exemplar of the effective solutions which the Stevensons designed and engineered.

That so many of their towers still fulfill their original purpose speaks volumes about the quality of their construction. Not least of the challenges they faced at Sumburgh Head was

[above] Robert Stevenson neither designed nor built Southerness lighthouse in the Nith Eastuary in Dumfries and Galloway, but he improved its efficiency with the addition of a silvered copper parabolic mirror and an Argand lamp between1812 and 1815.

[left] Sandsayre Pier, on the east coast of Shetland's Mainland, was built in 1854 by two of Robert Stevenson's sons, David and Thomas. Recently restored, it is used by the boat which takes visitors out to the island of Mousa.

[bottom] The 65ft (20m) high lighthouse on the Bass Rock, off the East Lothian coast, was built by David Alan Stevenson (1854–1938) and Charles Alexander Stevenson (1855–1950), and was first switched on in early November 1902.

getting building materials to the site – and part of the delay was caused by the loss of a ship and its crew carrying those materials.

Work started on building the lighthouse in 1819, and it was completed later than planned, in 1821. It became the first 'modern' building in a landscape which had first been colonised long before the Iron Age.

At Southerness, the beam could be seen for just over nine miles (14km). The Sumburgh beam, however, penetrated rather further due to the lighthouse's elevation on top of the cliff.

The contract for building the lighthouse was given to John Reid of Peterhead and it was planned that the light – which was an Argand oil lamp for the first few years – would be lit for the first time in autumn 1820.

The lamp – which had the power of six 'British Standard Candles', amplified by a highly polished mirror and beamed through a Fresnel lens – was invented in 1780 by Geneva scientist Aimé Argand. Six burning British Standard Candles only actually generated an equivalent amount of light to a modern two watt LED, so the importance of the reflector and lens in collecting and concentrating that light to make the lighthouse an effective warning beacon was immense. The simple oil lamp was replaced in 1868 with a high intensity paraffin lamp, which considerably increased the light's range but, perhaps surprisingly, electricity was not introduced for more than another century – until 1976 in fact.

Initially all did not go according to plan in the construction at Sumburgh and the lamp was not eventually lit until 15 January 1821. It had a limited range in the worst of Shetland weather. In those days, it was a stationary light rather than the rotating or flashing beam with which we are familiar today, and it would remain such for more than a century until a

rotating Fresnel lens array was introduced in 1914. An enhanced version of that lens system is still in use today.

Stevenson's specification for the tower was very detailed and, recognising just how severe the weather could be, recognised the need for a double-skinned wall with a total thickness of four feet to keep damp and cold out. It took longer to build than originally planned, but in the 200 years it has stood exposed to the elements, it has proved remarkably resilient, remaining completely damp-free inside.

Apart from the electricity, the lighthouse and its mechanism as seen today largely date from 1914. That was the point at which the revolving light assembly with its complex lens was installed. That lens is usually referred to as a 'Fresnel Lens', although Augustin Fresnel (1788-1827) did not invent it, but simply improved it.

Projecting the light further and further – as required by the increasing scale and frequency of shipping – meant manufacturing larger and larger lenses but, of course, the bigger the lens became, the heavier it would become, and the greater the energy required to rotate it. That's where the ingenuity of Fresnel came in. He was an engineer by profession, but was fascinated by light – and indeed, while still in his early twenties, had put forward some radical theories about the nature of light which, in time, would prove to have been absolutely correct. Several optical innovators before him had deduced that the bulk of a powerful lens could be significantly reduced by grinding a piece of glass down into a series of concentric zones which, collectively, would produce the same optical effect as a much bigger and heavier simple lens. The first was probably Count Buffon in France in the mid-18th century, and his design was improved by others, including Sir David Brewster in the early years of the 19th.

Fresnel's 1822 design took the idea a step further, using a lens built as a series of annular rings which intensified the light beam without the unwanted problem of spherical aberration, and he first applied the design to a lighthouse the following year.

The key to moving such a lens assembly was to float it on a bath of mercury, creating virtually frictionless movement. Indeed, a lighthouse lens could be rotated with just one finger if it was not constantly rotating – something it has to do as the lens focuses sunlight so precisely that were it not moving, it would create such a concentration of heat and light that fire would result.

The combination of rotating a complex array of lenses at a pre-determined speed, produced the consistent sequence of flashes which gave each lighthouse its signature.

The lens at Sumburgh was manufactured by Chance Brothers of Smethwick in Birmingham, and the associated mechanism was engineered and installed by James Dove & Company of Edinburgh, overseen by resident engineer David Alan Stevenson, grandson of Robert who would, either individually or with his brother Charles, actually oversee the design and construction of 90 lighthouses in his lifetime, five times as many as his more celebrated grandfather.

Fresnel lenses came in nine sizes, nine being the smallest, with a 'First Order' lens such as at Sumburgh being the largest. It measures over eight feet in diameter, has 26 reflectors, and the revolving lens and mechanism weighed an estimated three tons. By comparison, the lens on the former Muckle Roe light which was rebuilt at Sumburgh Head and greets visitors to the site, is described as 'Third Order Large', with a focal length of 20ins (500mm), about half that of the main lighthouse's lens.

The Stevenson family was at the leading edge of lighthouse design for almost 150 years. Their work built on the achievements of Smeaton and others, evolving lighthouse power and efficiency and saving countless lives along the way. The majority of their lighthouses are still operational today.

[top] One of James Dove's brass plaques from 1906.

[above] James Dove and built the fifty feet tall cast-iron octagonal Western lighthouse on Newhaven Pier near Edinburgh in 1869 but it was a further nine years before it was lit. It was decommissioned in 1930 and now, with coloured lights instead of a beacon, it is purely decorative.

[opposite page top] Inside a Fresnel lens array.

[middle] Corsewall Lighthouse, on the Ayrshire coast, was built by Robert Stevenson 1817 and is now a holiday let.

[bottom left] The 1853 Arnish Point lighthouse on Lewis, the last to be built by Alan Stevenson, was also the first to be built using prefabricated iron plates lined with wood.

[bottom right] Fair Isle North Lighthouse, built by David A. and Charles Stevenson, who also built the Fair Isle South light. Both were completed in 1892

JAMES WATT AND THE ENGINES
THAT CHANGED THE WORLD

James Watt's achievements are many and fascinating, and his importance in the history of engineering is immense. But separating fact from folklore can be challenging, so widely repeated are the often fanciful accounts of the birth of the steam engine. The importance of Watt and his partnership with Matthew Boulton in the history of steam power – and the value of preserving aspects of their heritage – was recognised in Victorian times, as the article which accompanied the photograph of the Smethwick Engine, top left, in the journal *The Engineer* made clear.

> 'During the present year—1898—this remarkable old engine, which has been regularly at work from the time of its erection to the current year, a period of, say 120 years, was removed to the canal company's station at Ocker Hill, Tipton, there to be re-erected and preserved as a relic of what can be done by good management when dealing with machinery of undoubted quality.
>
> It is worthy of note that the Birmingham Canal Navigations favoured Messrs. Boulton and Watt in 1777 with the order for this engine, and in 1898, or 120 years afterwards, the company has entrusted the same firm, Messrs. James Watt and Co., Soho, Smethwick, with the manufacture of two of their modern triple-expansion vertical engines, to be erected at the Walsall Pumping Station, having 240-horse power, and a pumping capacity of 12,713,000 gallons per day.'

The engine, now reassembled and restored to full working order in Birmingham's Thinktank Museum, is the oldest working steam engine in the world. No doubt, despite this published eulogy on its 'undoubted quality', even the columnist from *The Engineer* could never have imagined that the 'Smethwick Engine' would regularly be steamed again 125 years after he wrote his brief account of its history.

Born in Greenock, a small but thriving port on the Firth of Clyde, James Watt's childhood seems to have been spent quietly. His father, also James, was a man of many talents, becoming an accomplished instrument-maker – a talent his son inherited and, but for a quirk of fate, he might have spent his life making scientific instruments rather than introducing the world to the efficient use of steam power. According to Andrew Carnegie's 1903 biography, *James Watt*, the young James was fascinated by all things mechanical and mathematical from a very young age, and spent long hours dismantling and re-assembling mechanical toys to learn how they worked.

[above] A late 18th century portrait of James Watt.

[opposite page top] 'Constructed by Messrs. Boulton and Watt in the year 1777, the order being entered in the firm's books that year as a single-acting beam engine, with chains at each end of a wood beam, and having the steam cylinder 32in. diameter, with a stroke of 8ft., and erected at the canal company's pumping station at Rolfe Street, Smethwick.' From *The Engineer*, 15 July 1898.

[opposite page bottom] The Smethwick Engine is now exhibited and operated in Birmingham's Thinktank Museum.

connections with:

- James Brindley
- John Rennie
- William Murdoch
- Isambard Kingdom Brunel
- Sir William Armstrong

'We have another story of Watt's childhood that proclaims the coming man. Precocious children are said rarely to develop far in later years, but Watt was pre-eminently a precocious child, and of this several proofs are related. A friend looking at the child of six said to his father, "You ought to send your boy to a public school and not allow him to trifle away his time at home." "Look at how he is occupied before you condemn him," said the father. He was trying to solve a problem in geometry.'

The young Watt was not interested in any sport except fishing, and reportedly spent much time by the shores of the Clyde – in those days a very busy river with great sailing ships making their way up and down the river on their way to or from the great ports of the world.

In his late teens, his attempts to establish himself as an instrument-maker in Glasgow fell foul of the Guild of Hammermen, who refused to allow him to practise his skills as he had not undergone a formal apprenticeship. But to whom might he have been apprenticed? There were no qualified 'masters' in Scotland at the time – so like so many others at that time, James eventually had to leave Scotland and learn his skills in London.

As a teenager, however, deciding to leave home and make his own way in the world after the death of his mother, Agnes (née Muirhead), in 1753, James went to Glasgow to stay with his Aunt and Uncle Muirhead – his mother's brother John and his wife – where he was employed by a man Carnegie described as

> 'a kind of jack-of-all-trades, who called himself an "optician" and sold and mended spectacles, repaired fiddles, tuned spinets, made fishing rods and tackle, etc. Watt, as a devoted brother of the angle, was an adept at dressing trout and salmon flies, and handy at so many things that he proved most useful to his employer, but there was nothing to be learned by the ambitious youth.'

A fisherman by the banks of the Clyde a few miles upstream from Greenock. As Watt enjoyed the same hobby more than 250 years ago, activity on the river in front of him would have been relatively quiet, but the following years would see an increasing amount of traffic passing up and down as Glasgow started to evolve from a relatively small town into Scotland's premier industrial city.

According to some sources, it was his 'Aunt Muirhead' who later told the apocryphal, but often repeated, story of 15-year old James watching steam from the kettle boiling on the hearth as it condensed on a spoon, and later applying what he had observed to the 'invention' of the steam engine. Other sources ascribe the first account of the story to his cousin, Marion, John Muirhead's daughter, later Mrs. James Campbell, but as she was just 13 at the time, it is likely that the story was passed on at a later date by both mother and daughter.

While the story is unlikely to be true – although Andrew Carnegie was convinced of its authenticity – Watt clearly did use kettles in a number of the early experiments he carried out in order to ascertain the power and heating efficiency of steam.

In any event, Thomas Newcomen had already introduced considerable improvements into the simple steam pumping engines designed by Thomas Savery – which were already being used to drain a number of Cornish tin mines – 20 years before Watt was born. It was Newcomen who introduced the rocking beam to drive a piston, and it was he who first injected cold water directly into the steam chamber, condensing the steam quickly and thus creating a partial vacuum. It worked, but was fuel-hungry and not particularly efficient.

James Watt became involved in steam technology when working as a technician at Glasgow University and, when asked to repair a Newcomen steam engine, its shortcomings and inefficiency became immediately apparent to him. His 'improvements' hugely increased the engine's efficiency and, starting with the Smethwick Engine, his wide-ranging innovations in engine design would revolutionise production across the whole gamut of manufacturing and give birth to the modern industrial world.

Watt surveyed a possible route for a canal which would run from Crieff to join the meandering River Tay at the foot of Kinnoul Hill outside Perth, and which he believed would involve 'a distance of less than 20 miles, almost upon a dead level, which might be accomplished at an expense of about 10,000*l*' – the same figure he had suggested for the Monkland Canal. The plan foundered, only to be revived a few years later by John Rennie in 1805, his proposal suggesting that such a waterway 'would not only facilitate and render less expensive the conveyance of goods, wares, and merchandise between the towns of Dundee, Perth and Crieff and other places, and particularly the Highlands, but would also promote the improvement and cultivation of the adjacent country by conveyance of manure, and would be otherwise of much public advantage and utility; but the same cannot be effected without aid and authority of Parliament.' Like Watt's proposal, Rennie's was also eventually shelved and the projected Strathmore and Strathearn Canal was never built.

Recognising that Watt did not invent the steam engine in no way undermines the monumental nature of his achievements, as his skill lay in bringing together several separate ideas and combining them in an engine which more than trebled the efficiency and power of Newcomen's engine, thus reducing operating costs by up to 75%. The essential features of Watt's engine included greater precision in the boring of cylinders, better piston seals, enclosing the cylinder in a steam-heated jacket to stop it cooling mid-cycle, and the use of a separate condenser.

While Watt is renowned as the father of steam power, he also left his mark on the Scottish landscape in other ways. In his 1858 book *The Life of James Watt*, James Patrick Muirhead quoted Watt's own account of this period, gleaned from a surviving letter to a Dr. Small on 12 December 1769.

> "'I somehow or other", he says, "got into the good graces of our present magistracy, who have employed me in engineering for them, (as Mr. Smeaton terms it); among other things I have projected a canal to bring coals to the town;– for though coal is everywhere hereabout in plenty, and the very town stands upon it, yet measures have been taken by industrious people to monopolise it and raise its price 50 per cent, within these ten years. Now this canal is nine miles long, goes to a country full of level free coals of good quality, in the hands of proprietors who sell them at present at 6*d*. per cart of 7 cwt. at the pit. There is a valley from Glasgow to the place, but it has a rise of 166 feet perpendicular above our river; I therefore set that aside, and have found among the hills a passage, whereby a canal may come within a mile of the town without locks, from whence the coals can be brought on a waggon-way. This canal will cost 10,000*l* – is proposed 16 feet wide at the bottom, the boats 9 feet wide and 50 feet long, to draw 2^1/$_2$ feet water."'

His report on the Monkland Canal was published in 1769 under the title *A Scheme for Making a Navigable Canal from the City of Glasgow to the Monkland Coalierys*. As Muirhead noted, Watt was faced with somewhat of a dilemma over this project

> 'And, although "a determination that everything should yield to the engine" led him to refuse going to London with the bill for the Monkland canal, yet, after the Act for it had been obtained, and he was asked to superintend the

execution of the canal, he felt obliged not to refuse that request. "I had now a choice," he says, "whether to go on with the experiments on the engine, the event of which was uncertain, or to embrace an honourable and perhaps profitable employment, attended with less risk of want of success:– to carry into execution a canal projected by myself with much trouble, or to leave it to some other person that might not have entered into my views, and might have had an interest to expose my errors; (for everybody commits them in those cases.)"

He elected to see the canal project through to completion, but it is clear from his letter to Dr. Small that things were not proceeding as smoothly as he would have wished. The project had helped him appreciate how useful it would be having someone else to handle the business side, leaving him to the engineering – underlining the potential value of the sort of partnership he established with Matthew Boulton just a few years later.

[below] A short stretch of the Gartsherrie Branch of the Monkland Canal in Coatbridge. James Watt was responsible for surveying the main route of the canal in 1769. The branch dates from 1826.

[bottom] At the beginning of the 20th century the same location was the site of the massive Summerlee Iron Works, fragments of which still survive. Fittingly, this once important site is now home to the Summerlee Museum of Scottish Industrial Life.

"'Besides, I have a wife and children, and saw myself growing gray without having any settled way of providing for them. There were other circumstances that moved me not less powerfully to accept the offer; which I did; though at the same time I resolved not to drop the engine, but to prosecute it the first time I could spare. Nothing is more contrary to my disposition than bustling and bargaining with mankind:– yet that is the life I now constantly lead. Use and exertion render it rather more tolerable than it was at first, but it is still disagreeable. I am also in a constant fear that my want of experience may betray me into some scrape, or that I shall be imposed upon by the workmen, both which I take all the care my nature allows to prevent. I have been tolerably lucky yet; I have cut some more than a mile of canal, besides a most confounded gash in a hill, and made a bridge and some tunnels, for all of which I think I am within the estimate, notwithstanding the soil has been of the very hardest, being a black or red clay engrained with stones.'"

It is reassuring that Watt felt the same fears and pressures which beset us all. Worrying that his 'want of experience may betray me' – today known as 'imposter syndrome' and experienced by many – is a concern to which we can all relate. History has chosen to remember Watt as a much more confident man. The letter also discussed the scale of the project, and his motivation for continuing

'I have for managing the canal 200*l.* per annum; I bestow upon it generally about three or four days a week, during which time I am commonly very busy, as I have above 150 men at work, and only one overseer under me, beside the undertakers, who are mere tyros, and require constant watching. The remainder of my time is taken up partly by head-aches and other bad health, and partly by consultations on various subjects, of which I have more than I am able to answer, and people pay me pretty well. In short, I want little but health and vigour to make money as fast as is fit.'

By November 1772, with seven of the planned 12 miles completed and the original £10,000 budget completely spent, Watt was in some despair, painting a picture of himself to Dr. Small which few today would recognise as being realistic.'

> 'Remember also I have no great experience and am not enterprising, seldom choosing to attempt things that are both great and new;'

Muirhead was no more convinced by this self-effacement than historians are today,

> 'The names of the other engineers with whose Reports on the same subject that of Watt has thus been associated, viz. Smeaton, Golborne, Rennie, Whidbey, Clark, Hartley and Walker, bear conspicuous testimony to the advancement which Mr. Watt was now rapidly attaining in such pursuits, as well as to the clear-sighted discernment and public spirit of those who employed him.'

This woodcut purports to illustrate the moment when the 15-year old James Watt 'discovered' the power of steam. It was used on the cover of *The Illustrated London News* on 23 September 1867. The woodcut was based on a painting by Marcus Stone RA (1840–1921) exhibited at 'the French Gallery' in London.

That list referred to John Smeaton, John Golborne, John Rennie, Joseph Whidbey, William Tierney Clark, James Walker, and perhaps the surveyor and bridge-builder Bernard Hartley, father of the renowned dock engineer Jesse Hartley who in 1846 would build the first fully enclosed dock – the Royal Albert Dock in Liverpool. Today, while some are world famous, others have faded from public awareness.

Throughout this period of his life, while his civil engineering projects took up much of his time and put great strain on his health, Watt still found time to contnue work on his 'improved' engine, aided by the sponsorship of John Roebuck – who had invested heavily in the engine in return for two-thirds of the income expected from it.

When Roebuck declared bankruptcy in 1773, all Watt's efforts might have come to a standstill but for the establishment of what would become his pivotal partnership with Matthew Boulton. Samuel Smiles in his book *Industrial Biography: Iron Workers and Tool Makers* recalled Roebuck's decision

> 'At the same time, he transferred to Mr. Boulton of Soho his entire interest in Watt's steam-engine, the value of which, by the way, was thought so small that it was not even included among the assets; Roebuck's creditors not estimating it as worth one farthing. Watt sincerely deplored his partner's misfortunes, but could not help him. "He has been a most sincere and generous friend," said Watt, "and is a truly worthy man." And again, "My heart bleeds for him, but I can do nothing to help him: I have stuck by him till I have much hurt myself; I can do so no longer; my family calls for my care to provide for them."'

Whatever emotions he may have felt, developing the partnership with Boulton would free him from the dealing with the time-consuming business details of his activities, allowing him to devote himself more fully to innovation. Within just two years of establishing the partnership, their first large engine was built and operating on James Brindley's Birmingham Canal. It was an innovative engine in many respects, and not just because of its size and power.

One of the perennial problems with many canals is the quantity of water which is used [right] In Watt's earliest rotative engine, the up-and-down action of the beam was converted into a rotative motion by the 'sun-and-planet' gears around the central shaft of the flywheel, originated by William Murdoch. In 1878, 97 years after he built it, Watt's 1781 engine was illustrated in *The Development of the Modern Steam Engine by James Watt and his contemporaries* by Robert Henry Thurston. Very few of the ideas in this engine were original to Watt – some of them needed to be licensed from their patent holders – but Watt brought them all together in a single machine.

[below] This early Boulton and Watt engine with its wooden beam and frame was published in 1915 as a cigarette card by W.D. & H.O. Wills, No.12 in their series of 'Famous Inventions'. Portraits of Watt also featured on cards from Ogdens and Arbath, and Mitchell's of Linlithgow pre-1901.

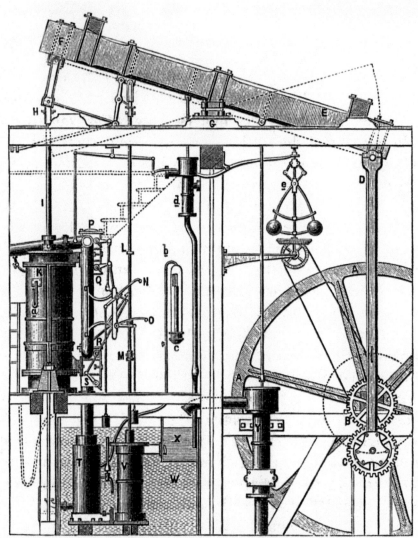

FIG. 27.—Watt's Engine, 1781.

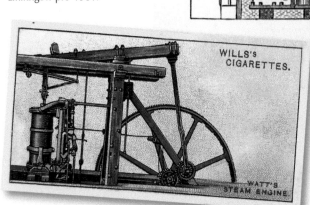

WILLS'S CIGARETTES.

WATT'S STEAM ENGINE.

– and lost – each time a boat descends through a lock – a problem exacerbated by a lack of feeder streams to keep levels topped up. It was thus central to the concept of what is now referred to as the 'Smethwick Engine' that the water lost at the foot of the flight of locks could be pumped back up to the top where it could be re-introduced into the system.

That simple innovation considerably reduced the amount of water needed to keep the locks filled – important in an area where other industrial demands on the available water supply were increasing.

The engine was able pump the equivalent of 1,500 buckets of water per minute – about one ton – up to the top of the flight of locks, giving the flight the capacity to handle up to 250 boats per week, and that was considered at the time to be more than sufficient to cope with likely future demand.

But demand rapidly exceeded the capacity of the locks, and within little more than a decade – by 1790 – a new cutting had been dug under the supervision of John Smeaton, taking the top three locks on either side of the summit out of the equation. At around the same time, the engine, was fitted with a larger pump to further increase capacity, and then had to raise water by just half the original height – 20 feet instead of 40.

When it was withdrawn in 1891 after 112 years service, its importance in the history of steam power was recognised immediately, and it was saved for preservation, first in a building in Tipton, then in the original Birmingham Science Museum and, finally, in the current Thinktank Museum after being painstakingly restored back to its 1850 configuration.

As the engine is configured today, the pump dates from an upgrade in 1790, while the piston and cylinder both date from the 1850 rebuild. Remarkably, throughout its long life – now more than 240 years – the engine has retained its original wooden beam. The durability of the wooden beam may have been considerable, but its days were numbered as the manufacture of large cast iron alternatives became a practical possibility. Boulton and Watt were pioneers of that new technology.

Watt was, understandably, highly protective of the unique features of his engines, and worded his patents very carefully to discourage others from duplicating his ideas. His primacy was also covered by the 1775 Steam Engine Act, which granted him extensive additional protection by Act of Parliament, triggering a series of patents by other inventors – amongst them Jonathan Hornblower and William Blakey – to develop engines which were to challenge his dominance.

Detail of the few metal components in the 1777 Smethwick engine. Its frame and beam were all made of wood.

Despite increased competition, Boulton and Watt's grip on the market was immense, their engines being widely used across Britain and exported worldwide.

The oldest-surviving rotative engine by Boulton and Watt was installed in 1785 in the Whitbread Brewery in London, and is now preserved in the Powerhouse Museum in Sydney, Australia, displayed as it looked around 1830, after two major rebuilds, one of which had involved replacing its second wooden beam with a cast iron one.

In the early years of the 19th century, the use of structural cast iron in mills and factories was becoming widespread, while the Boulton and Watt engines being installed a number of those factories were still predominantly built of wood.

Perhaps surprisingly, they were not at the forefront of the transition to cast iron, and it was not until the early years of the 19th century that they started developing their own distinctive range of iron beams. By 1819, that range extended to 16 different sizes of beams from just under 7ft (2m) in length to the massive beam at Crofton Pumping Station which measures nearly 27ft (8m).

One of the first engines of the 'new generation' was supplied in 1801 to the Carron Iron Works near Falkirk in Scotland, and the Boulton and Watt archive lists a 56hp (41.7kW) engine with an iron beam being ordered in January 1801 for J. W. & B. Bottfield's iron rolling mill in Coalbrookdale.

[right] This 1819 illustration shows the range of cast iron beams then available from Boulton and Watt. The beam at the top of the diagram is of the size and design which survives on the 1812 engine at John Rennie's Crofton Pumping Station on the Kennet and Avon Canal. The 26ft 8ins (8m) long cast iron beam weighs more than six tons. The engine is typical Boulton & Watt design of the period with parallel motion linkages at each end of the beam, and a separate condenser. It has a 42¼ ins (1073 cm) cylinder, and a 7ft (2.1 m) stroke. The fourth beam, measuring nine feet three inches from the fulcrum to the cross head pins, is the size and design used at Rennie's Claverton Pumping Station, also on the Kennet and Avon – see opposite. (*Library of Birmingham, MS 3147/5/1378*)

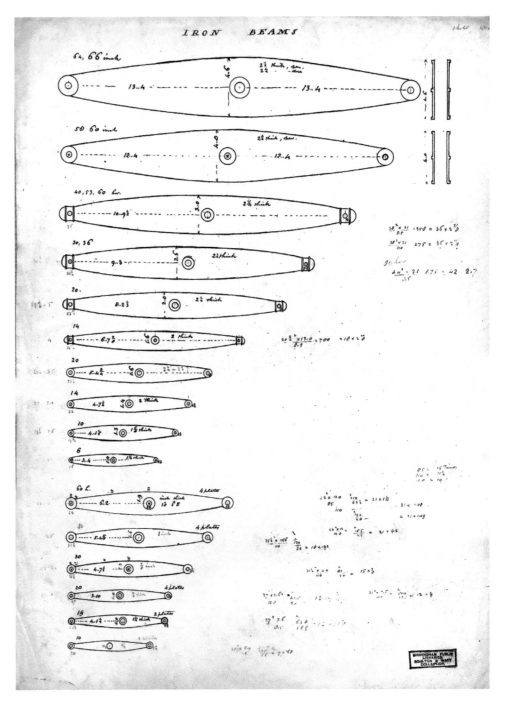

A 43hp (31.6kW) engine with an iron beam was built for the Chelsea Waterworks and installed at their Pimlico pumping station in 1803. It had a 48ins diameter cylinder and a stroke of eight feet, raising water from an aquifer more than 120 feet underground. It was one of more than 500 pumping engines Boulton and Watt had built by that time. All had been sold on the basis of their improved efficiency over other manufacturers' engines, and the huge savings in the cost of fuel which they delivered.

An engine with an output of 36hp (26.8kW), sun and planet gears, a 30.75ins diameter cylinder and a stroke of six feet, was supplied in December 1805 to William Balston's Springfield paper mill in Maidstone, Kent and was used to drive a 'Hollander' rag stamper which was to break down the cotton rags into pulp. It worked for around 80 years, being

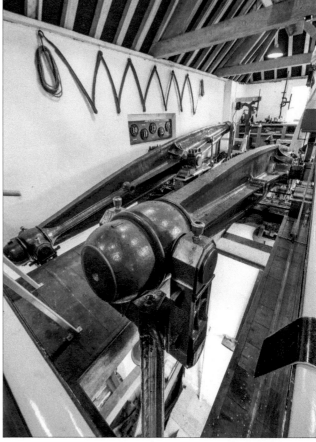

mothballed in the late 1880s and while the mill has since been demolished and the site redeveloped for housing, the beam survives, sitting on a plinth at the edge of the housing estate – *see page 59*.

This was a period of massive change for Boulton and Watt – they were still sub-contracting out some of the components of their engines in the closing years of the 18th century, a situation which was being phased out as the capacity of the new Soho Works expanded.

Work on building the new Soho Foundry had started in 1795 on a site specially chosen for accessibility, a mile from Matthew Boulton's original Soho Manufactory. The new works comprised a foundry, a forge, a smithy, turning lathes, and other workshops.

The Birmingham and Wolverhampton Canal was adjacent to the site, thus offering Boulton and Watt the ability to ship engine parts anywhere across the country along the expanding canal network. A new dock was excavated alongside the canal for the loading and unloading of machinery on to barges. The canals were ideal for shipping such heavy engines and parts.

Within a decade of the establishment of Soho, wooden framing and wooden beams had largely been phased out in favour of cast iron and the company also found itself developing a lucrative trade in upgrading existing engines, one example being the twelve horse-power engine with sun and planet gear built for the Enderby Brothers – Charles, Samuel and George – and installed in their whale oil mill near St Paul's Wharf in London. The original drawings for that engine were dated May 1801 but the engine was extensively upgraded and rebuilt in 1813, the original wooden beam being replaced by an iron one.

By 1801, William Murdoch had been appointed as Chief Engineer, the two founders having passed control of the business to their sons, James Watt Junior and Matthew Robinson

[above left] Crofton Pumping Station on the Kennet and Avon Canal is unique – still with one of its original beam engines intact, and capable of doing the job for which it was built. When the 1812-built Boulton & Watt beam engine was installed, it cost the canal company £2,040 including transportation and installation costs. When it and its 1846 Harvey & Company partner are in steam, the modern electric pumps are shut down, and the beam engines resume the task for which they were installed.

[above] No records have yet been found of the two Boulton and Watt beams on Rennie's 1815-built water-powered pumping engine at Claverton on the Kennet and Avon Canal being supplied directly by the makers when it was built.

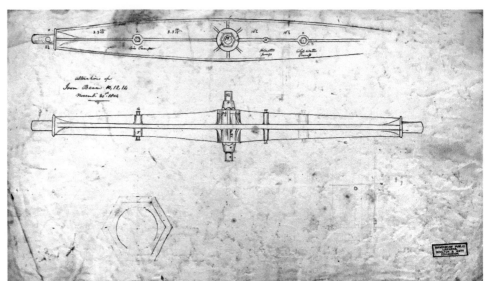

[above] Looking up towards the beam on the James Watt & Company engine at Eastney Pumping Station in Hampshire.

[top right] Detail from a working sketch in the Boulton and Watt Archive showing 'alterations' to 'Iron Beam 10, 12, 14' an early design, dated 30 November 1804. The pivot point to the left of the fulcrum connected to the air pump which drew water through the condenser and exhausted any air in the system. The right pivot connected to the water feed pump, while the inner pivot point right of centre evacuated the cooling water from the condenser. (*Library of Birmingham, MS 3147/5/1378*)

[middle right] An identical beam from a production drawing dated October 1824. (*Library of Birmingham, MS 3147/5/1378*)

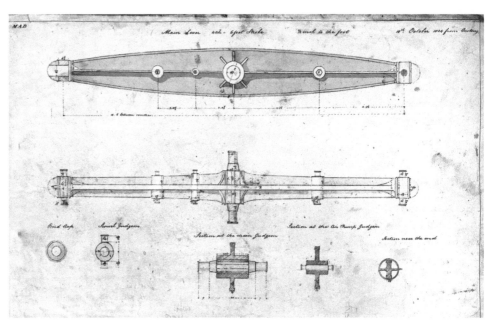

Boulton, in 1800, the year when Watt's original patents had expired, opening the market to be further developed by other engine builders.

Thus it was Murdoch who supervised the development of Phillips, Wood and Lee's new Salford Twist Company mill in Chapel Street, Salford, Lancashire – a seven-storey 'fireproof' spinning mill, the framing of which was entirely made of cast-iron. That structural design would become commonplace in the proliferation of cotton mills which opened in the following decades. Into the mill were installed the first of four Boulton and Watt engines, and each floor was entirely fitted with gas lighting.

The initial 50 gas lamps which Murdoch installed, eventually rose to an estimated 900. The gasworks, of course, was built by Boulton and Watt.

Despite Watt's achievements, the real future potential of the steam engine was only realised later in the 19th century, not with the simple double-acting single-cylinder engine, but in multi-cylinder giants which took their operating efficiency far beyond anything Watt had imagined.

Watt had certainly brought about huge improvements in efficiency with his precision-bored cylinders – using John Wilkinson's giant boring lathes – but there were even greater efficiencies to be harnessed. The key lay in the use of high-pressure steam – something which Watt had declared would be grossly unsafe – and engine designs which exhausted as little of the generated steam as possible. Improvements in boiler design by William Fairbairn and others would resolve the safety issues.

Matthew Boulton died in 1809, James Watt ten years later, before the true scale of their achievements could be realistically assessed. Boulton's death touched Watt deeply. To Boulton's son, Matthew Robinson Boulton, he wrote a moving tribute to his friend and business partner of almost 35 years. The letter was reproduced in Samuel Smiles' *Lives of Boulton and Watt, Principally from the Original Soho MMS*, published by John Murray in 1865. In a lengthy letter of condolence, he summed up his relationship with Boulton in a succinct paragraph.

> 'Few men have possessed his abilities, and still fewer have exerted them as he has done; and if to these we add his urbanity, his generosity, and his affection to his friends, we shall make up a character rarely to be equalled. Such was the friend we have lost, and of whose affection we have reason to be proud, as you have to be the son of such a father.

Smiles' almost evangelical support for Watt and his achievements found eloquent form, if a little over-generously, in a quotation from one of his contemporaries – the French engineer and politician Martial Bataille – in the closing pages of the book:

> 'Watt's engine was not an invention merely—it might almost be called a creation. "The part he played," says M, Bataille, "in the mechanical application of the force of steam can only be compared with that of Newton in astronomy, and of Shakespeare in poetry. And is not invention the poetry of science? It is only when we compare Watt with other mechanicians that we are struck by his immense superiority,—when we compare him, for example, with Smeaton, who was, perhaps, after him, the man who had advanced the farthest in industrial mechanism. Smeaton began, about the same time as Watt, his enquiries as to the best means of improving the steam engine. He worked long and patiently, but in an entirely technical spirit. While he was working out his improvements, Watt had drawn forth, from his fertile imagination all those brilliant inventions to which we owe the effective working steam-engine. In a word, Smeaton knew how to improve, but Watt knew how to create."

James Watt, Matthew Boulton and William Murdoch were all buried in St. Mary's churchyard in Handsworth, Birmingham; Watt and Boulton are also commemorated with memorials in Westminster Abbey. In addition, the achievements of all three are celebrated in "The Golden Boys" statue in Birmingham's Centenary Square, created in 1956 by William Bloye (1890–1975).

[above left] The 1805 beam from Springfield Paper Mill in Maidstone, Kent, is preserved in a modern housing estate. It is believed to be the oldest surviving cast-iron beam made by Boulton and Watt and matches the drawings opposite. *(courtesy of Bal Bhelay, Redrow Homes South East)*

[above] From *Steamships and Their Story*, published by Cassell and Company in 1910, a photograph of the simple single-cylinder engine from the 1812-built steamship, PS *Comet*. Towards the end of his life, Watt was taken for a sail in Henry Bell's pioneering vessel, and saw this engine in operation.

A working late-Victorian gas lamp, installed outside the Fakenham Gasworks Museum in Norfolk, the only 'town gas' works preserved in England.

WILLIAM MURDOCH'S GASLIGHT

In the 18th century, a new nickname was coined for the area around Redruth in Cornwall, one of the most mineral-rich areas of Britain, which claimed that it was 'the richest square mile on earth'. It could equally have been applied to several other square miles in west Cornwall.

The mining industry employed thousands of men in hundreds of mines, some of the biggest of them with 20 or 40 miles of tunnels deep below ground. Huge quantities of tin, copper and lead have been excavated from mines across the area for 2000 years at least, but most spectacularly in the 18th and 19th centuries.

Copper-rich, and then tin-rich, ore was to be found across Cornwall, and mines were dug deeper and deeper to retrieve it. Deeper mines invariably meant wetter mines, with the attendant need for reliable – and much more efficient – pumps to keep them drained. Such pumps had to move huge quantities of water round the clock if the profitability of those mines was to be maintained – especially in those areas where they were dug far out beneath the sea.

The great steam pioneer Thomas Newcomen, later followed by James Watt, recognised that the mines of Cornwall were just the places where improving the efficiency of the pumps could bring huge commercial benefits. The remains of the engine houses which kept them operational can still be seen all over the county.

By the late 18th century, hundreds of steam engines were being used to drain mines. In introducing them, Boulton and Watt came up with a novel business plan – they would

[above] An engraving of William Murdoch, or Murdock, in middle age, based on one of two portraits of him painted between 1823 and 1827 by the Scottish portraitist John Graham-Gilbert (1794–1866). One of the paintings is titled 'William Murdock', the other 'William Murdoch'. Graham-Gilbert also painted a portrait of James Watt, but apparently not one of Matthew Boulton.

[below left] Ruined engine houses, with which Murdoch would have been familiar, can still be found all over Cornwall. There are no accessible records, however, of whose engines were originally installed in these engine houses at Towanwroath near St. Agnes.

connections with:
- John Rennie
- James Watt
- William Fairbairn

Dolcoath mine's 'man-engine' in 1893, a Woodburytype from an image by photographer John Charles Burrow. The 'engine' took the miners to and from the working level. It had a series of 'steps' fitted to rods driven from the engine, using the motion of the beam to raise and lower those stepped platforms up and down the shaft. The miner rode up one 'level' on the upward action, stepped off on to a platform and waited for the next 'step' While quite dangerous, it was a lot safer and faster than climbing vertical ladders when exhausted.

lease the engines to the mine owners, on contracts which included servicing. Thus William Murdoch (1754–1839) was initially dispatched to Cornwall to install engines and subsequently to keep them operational – of paramount importance if income from the 'rental' system on which Boulton and Watt's business plan had been developed was to be sustained.

Throughout his lifetime, he was variously referred to as Murdock, or Murdoch but it is clear from his own writings, and the patents taken out in his name, that he used 'Murdock' suggesting that may have been his personal preference. That, however, has not stopped a continuing controversy over which is the correct spelling. More that 50 years after his death, a letter appeared in *The Engineer* berating those who chose to use the 'Murdock' spelling.

'SIR, In view of the fact that your correspondent "R.B.P. – has intimated in his first letter – is engaged in writing a sketch of the life of William Murdoch, I cannot help expressing regret that, in spite of his acknowledging that Murdoch is the correct way of spelling the name, he deliberately contemplates dropping the "h" and substituting "k" in writing of the inventor. I am all the more led to call attention to this matter on account of the fact that in a small memoir of Murdoch, just published, under the auspices, or at the request of the British Association of Gas Manufacturers... ...the writer – who is a Mr. A. Murdoch and a relative of the inventor – throughout the first half of the volume spells the name Murdoch, and in the latter half Murdock...'

Murdoch – for that was his father's surname, and the name he used well into his professional career – was born in Bello Cottage on James Boswell's estate in Ayrshire, where his father, John, was a millwright and miller, and a pioneer in the safe use of iron gearing in his mill's machinery. He rented Bello Mill from the estate in 1754, moving into the cottage just before William was born.

In the 1760s he had the mill's new machinery cast at the recently-established Carron Foundry near Falkirk, turning the water-mill into the most technically-advanced mill in Scotland. Also in the 1760s, the Carron company was chosen to manufacture some of the cast-iron parts for James Watt's early steam engines. Perhaps that is where and how the young William first became interested in the steam engines which subsequently to occupied much of his life.

From a very early age, he had been fascinated by steam power and the dawn of the machine age so, at just 23 years of age, he left Scotland and made his way to Boulton and Watt's Soho Works seeking employment – a decision which would ultimately define the direction of his entire professional life.

His initial role with Boulton and Watt was as an engine erector and he had an enviable reputation for being able to get maximum power out of their engines for the minimum use of fuel – very important as, there being no coal mines in Cornwall, all the fuel had to be 'imported' from coalfields in Somerset and Gloucestershire. He thus played a pivotal role in the development of the steam engine, introducing some key innovations for which he was not credited at the time.

It is estimated that there were more than 600 steam pumps in use at times of peak production in Cornwall's larger mines. Many smaller mines continued to use horse-driven pumps – often tended by young boys who kept the horse moving – as a cheaper alternative.

For those electing to use steam pumps, in times of high mineral prices the deal was cost-effective, but in times of slump, with rental payments still to be met, and Boulton and Watt 200 miles away, Murdoch as the man on the spot, bore the brunt of the mine-owners' disquiet.

At its peak in the mid-19th century almost one fifth of an estimated workforce of 30,000 to 40,000 across the industry were women. While they stayed on the surface, the men made the long journey below ground every day. Before the advent of winding engines, that involved a perilous climb down a succession of near-vertical ladders in pitch darkness, with a fall of hundreds of feet awaiting anyone who lost their footing. Coming back up exhausted after a hard day's work with a pick and shovel must have been even more perilous.

[above] Mining students at work underground. Although the idea was first proposed ten years before Murdoch's death, the Camborne School of Mines was not established until 1888. Now part of the University of Exeter, it is still one of the leading institutions for training mining engineers. Photographed in 1893 by J. C. Burrow.

The author Wilkie Collins was invited down a copper mine in the 1850s, and descended around 240 feet before, perhaps wisely, deciding that the view was not going to improve if he went any lower! He included an account of his experience in his 1851 book *Rambles beyond railways or notes in Cornwall taken a-foot*.

By the second half of the 19th century when the deepest shafts were being dug, applying some of the power of the steam beam pumping engine to drive a 'man engine' (*see opposite page*) speeded up the miners' descent and ascent. A true genius may not be the one who makes an initial discovery, or series of discoveries, but the one who successfully applies those discoveries to the benefit of progress. The 'man-engine' not only reduced journey times to and from the work face, it significantly increased productivity.

[below] Murdoch designed and built this oak-framed 20-ton hand-cranked crane for the Soho Foundry, c.1800. The oak centre-post was replaced with a cast iron one in the 1870s, and the steam engine was added in the 1880s. The crane remained in use for over 100 years.

Just as James Watt brought together the inventions of a number of people in the improvement and refinement of the steam engine, so Murdoch himself drew upon the discoveries of others in the development of gas lighting, the discovery for which he is best remembered. According to a retrospective article about his achievements which appeared in the journal *The Engineer* on 4 March 1870:

'William Murdoch, one of Watt's chief and valued assistants in Cornwall, whose name is well known in connection with gas lighting, was much esteemed by the Cornish miners; he contributed very much towards the improvement of the pumping engine by improving the details as experience dictated. It is said that a party of Cornish mine agents were the first who saw a room illuminated with gas, and drank success to the new invention and its bold projector. He is also said to have delighted in astonishing the miners whom he sometimes met in the dark lanes, by carrying a bladder of gas under his arm and letting it out in lighted jets.'

An early 20th century triple-mantle gas light in the Fakenham Gasworks Museum in Norfolk.

[opposite top] Murdoch's 'sun and planet' gearing on an 1808 Boulton and Watt beam engine, illustrated as Plate XIX in John Farey's book *A Treatise on the Steam Engine* published in 1827. This engine is described as 'Mr. Watt's Rotative Engine of thirty-six horse-power constructed by Messrs. Boulton, Watt & Co. in 1808'.

[opposite middle] A postcard of 'Murdock House, Redruth' possibly dating from the 1920s, but using a much earlier photograph. The house, which has a very different appearance today, had no name when William carried out his gas experiments there.

[opposite bottom] The plaque commemorating William Murdoch's years in Redruth is affixed to the gable of Murdock House, and can just be seen in the 1920s postcard. It was placed there by the Birmingham engineers, Tangye Brothers Ltd., who built some of the largest and finest industrial steam engines of the 19th century.

Just as there is surviving evidence that Murdoch was responsible for several of the inventions attributed to James Watt – including the 'Sun and Planet' gearing which turned the vertical motion of the beam engine into rotative motion, without infringing existing patents – it is a matter of historical record that the recognition of the illuminating properties of burning gas predates Murdoch's involvement by more than a century. The issue of The Royal Society's journal *Philosophical Transactions* for 3 June 1667 included a piece titled *Description of a Well, and Earth in Lancashire, taking Fire by a Candle approached to it,* 'imparted by that Ingenious and Worthy Gentleman Thomas Shirley of Wigan reported that

'About the later end of February 1659, returning from a Journey to my house in Wigan, I was entertained with the relation of an odd spring, situated in one of Mr. *Hawkley's* Grounds (if I mistake not) about a mile from the Town, in that Road which leads to Warrington and Chester. The people of this Town did confidently affirm, that the Water of this Spring did burn like Oyle; into which error they suffered themselves to fall for want of a due examination of the following particulars.

For when we came to the said Spring (being five or six in company together) and applyed a lighted Candle to the surface of the Water, 'tis true there was suddenly a large flame produced, which burned vigorously; at the sight of which they all began to laugh at me for denying, what they had positively asserted; but I, who did not think my self confuted by a laughter grounded upon inadvertancy, began to examine what I saw.....'

What Shirley and his associates were setting alight was methane gas escaping from the huge resources of coal which lay beneath the Wigan area and which, over the centuries of coal-mining activity in the area, was the cause of numerous explosions and loss of life in the years before the development of miners' safety lamps. In that same journal in 1739, Dr. John Clayton recounted how he had acquired a quantity of coal from close by the 'burning brook' or 'burning well' and that he had

'distilled it in a retort in an open fire. At first there came over only phlegm, afterwards a black oil, and then likewise a spirit arose which I could noways condense. I observed that the spirit which issued out caught fire at the flame of the candle and continued burning with violence as it issued out, in a stream which I blew out, and relighted again, alternatively, for several times.'

What Clayton described as 'spirit' we reconise as 'gas' or 'coal gas' – terms which had not yet entered the vocabulary. He gathered his 'spirit' in a bladder and marvelled that even days later, when he let some out, it burned with the same intensity. He later reported that he often took one of his bladders, pricked a hole in it, ignited the 'spirit', and created a jet of brilliant light 'when I have a mind to divert strangers'. Can Murdoch's replication of Clayton's party-piece be taken as evidence that he was aware of Clayton's experiments?

A story recounted to William Matthews by one of Murdoch's acquaintances – the eminent engineer William Fairbairn (*qv*) – of a journey to Fairbairn's Manchester house one night, and published in Matthews' book *A Historical Sketch of the Origins and progress of Gas Lighting*, published by Simpkin and Marshall of London in 1832, underlines the likelihood.

'It was a dark winter's night and how to reach the house over such bad roads was a question not easily solved. Mr. Murdoch, however, fruitful in resource, went to the gasworks then established in Manchester where he filled a bladder which he had with him and, placing it under his arm like a bagpipe, he discharged through the stem of an old tobacco pipe a stream of gas which enabled us to walk in safety to Medlock Bank [Fairbairn's Manchester home].'

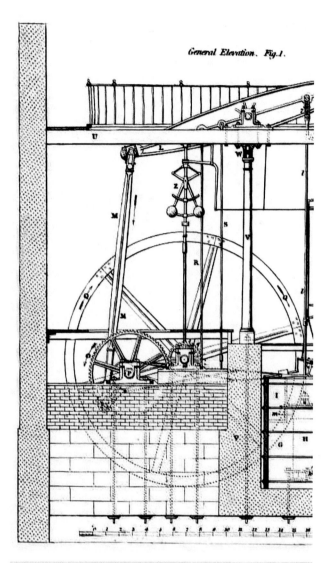

General Elevation. Fig.1.

Murdoch claimed he had only become aware of the experiments of Thomas Shirley, Lord Dundonald and Dr. John Clayton after his own successful work at Redruth, but insisted, rightly, that he had been the first to recognise and exploit the commercial potential of gas as an illuminant.

His own account, published in the 1832 edition of Matthews' book, declared that at Soho

'I constructed an apparatus upon a larger scale, which during many successive nights was applied to the lighting of their principal building, and various new methods were practised of washing and purifying the gas... ...Since that period I have, under the sanction of Messrs. Boulton, Watt, and Co., extended the apparatus at Soho Works, so as to give light to all the principal shops, where it is in regular use to the exclusion of other artifical light.

While the idea of using gas for lighting may not have been originally his, to Murdoch must go recognition of the technical and commercial potential of the idea and the development of the technologies required to gather and store inflammable gases.

In recognition of his achievements, he had become only the third recipient of the the Royal Society's prestigious Rumford Gold Medal in 1808.

Others laid claim to the primacy of the idea of course, some well grounded, others less so, the most notable being Frederick Winsor, although having no engineering background, fought through the courts in his attempt to get recognition as the inventor of gaslighting and reap the not inconsiderable financial rewards which could be

Murdock House, Redruth

WILLIAM MURDOCK
LIVED IN THIS HOUSE
1782-1798
MADE THE FIRST LOCOMOTIVE
HERE AND TESTED IT IN 1784
INVENTED GAS-LIGHTING AND
USED IT IN THIS HOUSE IN 1792

accrued. He claimed to have installed the first gas street lights – in Pall Mall in 1807 – but Murdoch had lit Chapel Street in Salford in 1805 – outside Phillips, Wood & Lee's mill, which he had already lit by gas inside.

A more important figure was Winsor's onetime partner, the Manchester-born engineer Samuel Clegg (1781–1861) who had served his apprenticeship at Boulton and Watt and

Thomas Rowlandson's 1807 cartoon of how gas lights worked.

Gentleman "The Coals being steam'd produces tar or paint for the outside of Houses -- the Smoke passing thro' water is deprived of substance and burns as you see."

Irishman "Arragh honey, if this man bring fire thro water we shall soon have the Thames and the Liffey burnt down -- and all the pretty little herrings and whales burnt to cinders."

Bumpkin "Wauns, what a main pretty light it be: we have nothing like it in our Country."

Quaker "Aye, Friend, but it is all Vanity: what is this to the Inward Light?"

Street Girl "If this light is not put a stop to -- we must give up our business. We may as well shut up shop."

Client "True, my dear: not a dark corner to be got for love or money."

[opposite top left] The Soho 'Gasometer' as illustrated in *The Engineer* in 1923. Despite now having now been attributed to Murdoch for more than a century since David Brownlie's article on the Soho Works appeared, this was not Murdoch's original 1798 gasometer, but its replacement, built 50 years later in 1848, nine years after the inventor's death.

[opposite top right and bottom] The similarities between the Soho retort house and that at Fakenham Gasworks in Norfolk are remarkable.

A PEEP AT THE GAS LIGHTS IN PALL-MALL.

had actually been present, working as his assistant, during the period when Murdoch was developing his gasworks at Soho. He also went on to play a significant role in the evolution of both gasworks and gas lighting technology, being responsible, amongst other things, for the installation of gaslighting at Stonyhurst College in Lancashire in 1811.

Murdoch's first operational gasworks was, unsurprisingly, built at the Soho Works, and started production around 1798. It was a remarkable feat of design and engineering, establishing a template for hundreds of other gasworks across the country. Indeed, it was so remarkable that the journal *The Engineer* was still marvelling at it 125 years later. In the issue for 30 March 1923, David Brownlie, in describing the photograph of the retort house interior *(see opposite)*, wrote:

> 'one of the identical installations erected by Murdoch at the Soho Foundry, somewhere around 1798, and is an astonishing piece of work, with the retorts in settings of three, direct fired underneath, with the standpipes from each retort going to the hydraulic main across the top of the setting and special arrangements for cleaning the tar out of the pipes.'

While more usually referred to as a 'gas holder' in recent years, Murdoch chose the term 'gasometer' because his device, illustrated in the same issue, did much more than merely contain the gas. His 'gasometer' consisted of a floating vessel filled with gas, the gas being kept in by a water seal.

As gas was pumped into the vessel – actually one vessel inside another – the pressure of gas caused the outer vessel to rise, albeit assisted by pulleys when necessary. As the gas was used, the outer vessel would slowly sink, maintaining a relatively consistent pressure of gas as it did so.

William Fairbairn had only become acquainted with Murdoch late in the latter's life, during the construction of McConnel & Kennedy's new Sedgewick spinning mill in Ancoats, Manchester. This was explained in his autobiography, completed after his death by William Pole:

'It was during the progress of this work (in 1818) that I became acquainted with Mr. Murdoch, of Soho, a gentleman well known to science and to the public as the inventor of the D valve, the improver of the Cornish pumping-engine, and the author of illumination by carburetted hydrogen gas. Mr. Murdoch was at this time upwards of seventy years of age, tall and well-proportioned, with a most benevolent and intelligent expression of countenance. He was the oldest mechanical engineer of his day, and, exclusive of his discoveries in practical engineering, he contrived a variety of curious machines for compressing peat moss, when finely ground and pulverised, into the most beautiful medals, armlets, bracelets, and necklaces, which, under immense pressure, being stamped and brilliantly polished, had all the character and appearance of the finest marble.'

Fairbairn had been involved in engineering the shafting for the mills – a group of four mills erected between 1818 and 1820 and powered by a 54hp (40kW) Boulton and Watt engine. He did not mention a date for the introduction of those medals, but in the early years of the nineteenth century Murdoch would have been just one of a number of people exploring the idea of 'faux marble' made from organic materials which could be compressed and moulded.

Several inventors in the 1850s would achieve considerable success creating – and patenting – 'thermoplastics' based on compressed, moulded and polished wood pulp and shellac. These would be used to mould elaborate cases – know as 'Union Cases' – for photographic portraits. There is, however, no record of Murdoch having sought to patent his material.

Murdoch's design for a steam carriage did get as far as a working model and a draft patent which he had every intention of taking to the Patent Office in London and filing. Watt had suggested that such a vehicle was already, *de facto*, covered by one of *his* own patents, and sent [right] Samuel Clegg installed gas lighting and a gasworks in Stonyhurst College in Lancashire in 1811, the first school in the world to have such a modern system. The two on-site gasometers each had a capacity of 500 cubic feet of gas, and the system used dissolved lime to remove toxic hydrogen sulphide. During the planning of the installation, there was much discussion about how the pipework – or 'conductors' as they were referred to – might be concealed as 'that would not look well in our refectory, study places, etc.' This stereoscopic (3D) view of the interior of the college's refectory, taken in 1858 by the eminent photographer Roger Fenton, shows the four-arm 'rats-tail' burners suspended from the ceiling, with the supply pipework presumably concealed above the plasterwork.

[below] From the German encyclopaedia *Meyers Konversations-Lexikon,* one of Watt's low-pressure engines with Murdoch's long D slide valve at the left.

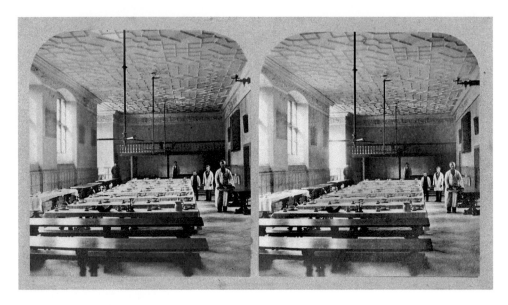

Boulton to intercept him *en route* and encourage him to drop the idea and concentrate on beam engines and gas lighting instead.

In the early years of this century, a group of enthusiasts built a full size version of the machine which, apparently, proved very challenging to drive. It is now in the Moseley Heritage Museum at Redruth and is steamed occasionally.

While he was an employee of Boulton and Watt, most of Murdoch's innovations were patented in James Watt's name, but there were exceptions – three patents in his own name were granted in 1791, 1799 and 1810 respectively, and while two of them were not directly connected with steam engines, the third definitely was.

The first of the three – No.1802 of 1791 – addressed the challenges of protecting timber – especially in ships' hulls – from the effects of submersion in waters. In creating his process, he employed raw materials whch were readily available near where he lived in Cornwall – an ore known locally as 'mundick', a form of pyrites rich in arsenic, zinc and sulphur which he heated in an oxygen-depleted retort, not dissimilar to that used in his gas experiments. The resulting yellow powder was to be used as a dyestuff or as the basis for a special paint for ships' bottoms. It was one of those patents, typical of the time, where far-reaching claims of primacy to inventions were made, often citing versions of the 'invention' based on conjecture rather than empirical support. Its title set the tone, claiming that he had

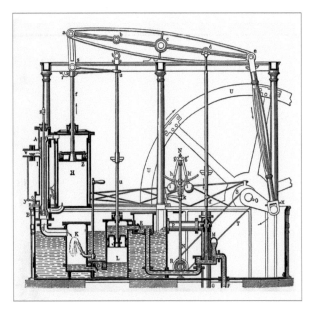

'after much labour, study, and expence, discovered "The art or Method of Making (from the same Materials and by Processes entirely New) Copperas, Vitriol, and Different Sorts of Dye or Dying Stuff, Paints, and Colours, and also a Composition for Preserving the Bottoms of all Kinds of Vessels, and all Wood required to be Immersed in Water, from Worms, Weeds, Barnacles, and every other Foulness which usually does or may adhere thereto."

His third patent – No.3292 of 1810 – was for a method of boring pipes and cutting cylinders out of solid stone, using manual power, a watermill or a steam engine. The methodology described was not widely taken up. For that patent, he described himself as 'William Murdock, of Soho Foundry in the parish of Harborne' for, with Watt and Boulton's sons, James Jnr. and Matthew Jnr., then operating the business, he had been appointed as Chief Engineer.

In between those two patents was the most interesting of the three – No.2340 of 1799, simply titled 'Steam Engines', which he filed while still based in Redruth. It contained new methods of boring cylinders and, more importantly, introduced a simplified arrangement of valves – with which he had apparently already experimented on his steam carriage. According to the specification, the innovations which he included in the patent simplified

'the construction of the steam valves or regulators and working gear in the improved engine of Mr. Watt's construction, called the double engine' [and] consists of connecting the upper and lower valves so as to be worked by one rod or spindle, and in making the stem or tube which connects them hollow, so as to serve for an eduction pipe to the upper end of the cylinder, by which means two valves are made to answer the purpose of the four used in Mr. Watt's double engine.'

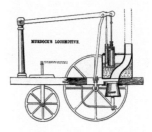

[above] The late Nigel Eden at the controls of the 'Murdoch Flyer', the full size reconstruction built by a local group of enthusiasts. It is now in the Moseley Heritage Museum, Redruth. (*Copyright unknown, from the Collection of Redruth Town Council.*)

[left] Murdoch's 1784 design for a steam-powered vehicle. He had been working on the project at his Redruth home for a number of years, creating a working model which was a source of local fascination as he gave it test runs in his garden and on local streets.

[below] A twin-mantle gas lamp at the entrance to Clacton-on-Sea Pier in Essex – a detail from a 'Photochrom' coloured photo-lithograph c.1896.

It became known as the long 'D slide valve' on account of its shape, and, while not endorsed by Watt, it became a feature, in various forms, of later designs of higher pressure engines introduced by other manufacturers. High pressure engines were another thing of which Watt disapproved. He had already persuaded Murdoch to abandon his oscillating engine design, and his steam vehicle.

Requiring Murdoch to permit so many of his other innovations to be patented in his master's name may have obscured just how much of the development of the steam engine he was responsible for. Their working relationship seems to have been strongly one-sided in Watt's favour, which makes Murdoch's professional loyalty remarkable. However, with hindsight, it can be argued that Murdoch's many contributions to the evolution of the steam engine considerably outweigh his role in the evolution of gaslight. Popular history, may have chosen to portray him as the loyal employee of Boulton and Watt, and eventually Chief Engineer, who invented gas lighting, but a more accurate description would be of a man who, as a result of his varied interests, left an enduring legacy across a number of industries. Precise details of his input into gaslight and steam power may be lacking, but his importance is beyond doubt.

THOMAS TELFORD, 'THE COLOSSUS OF ROADS'

The birth and christening of Thomas Telford (1757–1834, son of an Eskdale shepherd, were both registered as 'Thomas Telfer'. His father, John – known both as 'Telfer' and 'Telford' –, died when Thomas was three months old.

The Poet Laureate Robert Southey once described his good friend Telford as the 'Colossus of Roads', so important was his contribution to developing the nation's transport system. On a plaque on the bridge which crosses the river at Invermoriston on the north shore of Loch Ness, can be read the claim that this bridge was 'one of nearly a thousand built by Thomas Telford between 1803 and 1819' to improve the transport system of the Scottish Highlands – and if that is close to the truth, it's more than 60 bridges a year in Scotland alone.

According to Samuel Smiles, writing in his 1863 book *Industrial Biography*, Telford was very aware of the debt he owed to the vision of those engineers who preceded him. A significant number of his projects, as has already been mentioned, owed a great deal to the input of William Jessop. His own legacy of bridges across Britain, he openly acknowledged, would never have been possible without the vision, 80 years earlier, of the Darbys of Coalbrookdale who built the world's first iron bridge.

> 'While the work was in progress, Mr. Telford, the well-known engineer, carefully examined the bridge, and thus spoke of its condition at the time:— "The great improvement of erecting upon a navigable river a bridge of cast-iron of one arch only was first put in practice near Coalbrookdale. The bridge was executed in 1777 by Mr. Abraham Darby, and the ironwork is now quite as perfect as when it was first put up. Drawings of this bridge have long been before the public, and have been much and justly admired."

In a working life which spanned more than 50 years, Telford's works changed the face of Britain. He designed and oversaw the construction of thousands of miles of roads, some of the most imposing bridges of his day – including the Craigellachie Bridge opened just over 40 years after Darby's ground-breaking Iron Bridge – hundreds of miles of canals, major docks and harbours, and a host of little churches.

Despite his many achievements, however, Thomas Telford is less well-known than the likes of George Stephenson, Isambard Kingdom Brunel and other great engineers of the 18th and 19th centuries. Yet he is said to have designed, engineered those 1,000 bridges and many of

[above] A contemporary engraving of Thomas Telford aged about 53.

[opposite page] Opened in 1805 having taken ten years to construct. the eighteen-arch Pontcysyllte Aqueduct is 336 yards (307 m) long, and still carries narrowboats across the Dee valley. The longest canal aqueduct in Britain, at 126ft (38.4m) high, it is also the tallest in the world.

connections with:

- James Watt
- John Rennie
- William Jessop
- Isambard Kingdom Brunel
- The Railway Stephensons

[top] Dinham Bridge over the River Teme, engineered by Telford in 1823 to a design by John Straphen.

[above] Telford's tollhouse, seen beneath William Hazledine's chains on the approach to the Conwy Bridge.

[above right] The 326ft (99.3m) span of the Conwy suspension bridge, looking towards the imposing ruins of Conwy Castle.

[right] The Conwy bridges – Telford's 1826 suspension bridge with Fairbairn and Stephenson's 1849 tubular railway bridge alongside.

the industrial buildings which still stand around the country's smaller and now largely redundant docks.

Many of his warehouses have been converted into upmarket apartments, thus preserving important parts of the country's industrial and architectural heritage.

Almost wherever we look across the length and breadth of the country, from Berriedale on the north east coast of Caithness to the heart of London's docklands, the influence and genius of Thomas Telford can still be seen.

Telford has never been given the level of recognition which his legacy warrants. When Scotland was building new towns in the post-war years, neither the developers of Cumbernauld or Glenrothes considered naming their new town 'Telford'. That was left to the new town in Shropshire developed in the 1960s and 70s where Telford had been the appointed as that county's Surveyor of Public Works in 1787.

Several of Telford's bridges were engineering milestones in their day – amongst them are the great suspension bridge across the Menai Straits linking Wales and Anglesey, the smaller but equally important Conwy Suspension Bridge, the Pontcysllte Aqueduct near Llangollen, the elegant single-arch bridge over the Spey at Craigellachie, and his five-arch bridge across the River Tay at Dunkeld.

Three of his most spectacular bridges were constructed in North Wales – The Menai Bridge and the 1815-built Waterloo Bridge at Betwys-y-Coed, both on the main road from London to Holyhead, which was then a major port, and the Conwy bridge on the North Wales Coast road from Chester to Bangor.

The Holyhead road and its magnificent bridges saw coach journeys from London to Holyhead port reduced from more than 24 hours to less than nine.

The Menai Suspension Bridge opened first to much fanfare – on 30 January 1826 – with the Conwy bridge a few weeks later, making the Menai bridge the oldest wide-span suspension bridge in the world. It looks very different today, however, as it underwent a major renovation in 1938 which saw the original wrought-iron chains replaced by steel. In the early 19th century when the bridge was being built, Telford relied on the expertise of forge-master William Hazledine and the skills of his foundrymen at Plas Kynaston near Llangollen in North Wales to manufacture the wrought iron and cast iron chains and bolts.

The central span of the bridge, at 577ft (175.9m), was almost twice as long as the Conwy Bridge and it was clear at the design stage that much heavier and more robust chains would be needed to counter the impact of the severe winds which can pass through the Menai Straits.

[above] Ellesmere Port docks now house the National Waterways Museum. The 'dumb-boat' *Scorpio* was built as *Helena* by the Wigan Coal & Iron Company in 1890 to ship coal on the Leeds-Liverpool Canal.

[left] An Edwardian postcard of the Menai Bridge gives a clear view of the structure of the four sets of chains. The individual bars and links were transported from Plas Kynaston to the Welsh side of the bridge where the workforce assembled the chains on site before raising them into position over cast-iron saddles on top of each tower. The bridge's original wooden decking proved too light to stabilise the structure in side winds and was replaced in 1893 by a much heavier steel deck. The four sets of wrought-iron chains were replaced by two sets of steel chains in 1938.

Entrance to Menai Suspension Bridge.

[above] The bridge over the Tay at Dunkeld, completed in 1809, was Telford's longest bridge in Scotland.

[below] Craigellachie Bridge, the single span over the River Spey, was built between June 1815 and October 1816, at the cost of £8,200 pounds – about £1.5M at today's prices.

[below right] The cast-iron plaque on the south tower of the Craigellachie Bridge recalling that the ironwork was cast by William Hazledine at Plas Newyd in North Wales.

The design, therefore, called for three suspension chains – the additional one in the centre of the carriageway – supporting the wooden carriageway which spanned the strait 102ft (31m) above high water level, to enable the passage of tall-master ships beneath it.

Hazledine had initially established his foundry there in order to manufacture the cast-ironwork for Telford's and Jessop's 1000ft (300m) wide Pontcysyllte Aqueduct – listed as a World Heritage Site since 2009 – which still carries the Ellesmere Canal across the Dee valley which lies more than 120ft below it. That project had been the start of an enduring partnership, with ironwork from Plas Kynaston subsequently being used in a number of Telford's other great projects.

To speed up the manufacturing process, Hazledine had installed James Watt's recently developed beam engines in his foundry to lift strike hammers. These were not steam hammers in the later widely accepted sense as patented by James Nasmyth and others – rather they were simple tilt hammers in which the beam engine was used to lift the hammer head, allowing its own weight to power the fall. The high quality of Hazledine's output developed a strong bond between him and Telford, with Plas Kynaston iron being used on projects hundreds of miles from the foundry. The poet Robert Southey on seeing the Dunkeld bridge for the first time wrote,

> 'The bridge is one of Telford's works and one of the finest in Scotland. The Duke was at the expence, Government aiding him with 5000£. There are five arches, the dimensions of the five middle arches of Westminster Bridge; and besides these are two upon the land.'

[top left] The 'Bridge over the Atlantic' links the Scottish mainland with Seil Island. The single-arched, hump-backed Clachan Bridge spans Clachan Sound, seven miles south-west of Oban. Designed by Telford, it was built in 1792–93 by engineer Robert Mylne. Until the Skye Bridge was opened in 1995, this tiny bridge could claim to be the only one in Scotland to span the Atlantic Ocean.

[above] Telford's innovative Bannockburn bridge, completed in 1819. From an engineering point of view, it derives its strength in a remarkably similar way to the mid-14th century 'scissor arches' in Wells Cathedral in Somerset.

[above left] Tongland Bridge in south-west Scotland was opened in 1808, having taken four years to build. It cost £7,710. While the structure of the bridge was designed by Telford, the 'baronial' embellishment with castellated towers and parapet was the work of the artist Alexander Nasmyth whose son, James, would later be credited with the development of the steam hammer.

What fascinated Southey was the story of how the bridge had been built. Before work started, rubble washed down from the River Braan, a tributary of the Tay which joined the river just west of Dunkeld, had caused the Tay to change its channel, leaving a dry river bed running to the south of the ruined cathedral. So Telford had the unique opportunity to build a bridge on dry land, and 'by this means the building was carried on with greater ease, and at much less expense.' Once the bridge works were completed, the wash-down from the Braan was excavated away, and the Tay returned to its original course.

For the Craigellachie Bridge, with its 150-feet-wide span erected between 1812 and 1814, the ironwork was cast and forged at the Plas Kynaston and then shipped by barge across the Pontcysyllte Aqueduct and along the Ellesmere Canal – now known as the Llangollen Canal – and on to Chester. From there it was then loaded on to ships and carried by sea to Speymouth, between Lossiemouth and Buckie. That was years before the Caledonian Canal was completed – another Telford project – so the ships had to sail round the top of Scotland and back south to Spey Bay. From there the bridge sections were carried by horses and carts to the construction site for reassembly.

Such was the quality of both Hazledine's cast iron foundrywork and Telford's innovative design, that the bridge continued to carry the main road for almost 150 years, and now over 200 years old and recently refurbished, is still remarkably strong, albeit now only used by pedestrians and cyclists.

Telford's magnificent enclosed St. Katharine Dock in London was approved by Act of Parliament in 1825 and took three years to build, opening in 1828. Steam engines designed and built by Boulton & Watt – the company by that time controlled by the sons of the founders – kept the water level in the basins constant irrespective of the state of the tide outside the lock gates. The layout and cargo-handling facilities within the enclosed docks inspired later dock architects and engineers including Jesse Hartley when he was designing Liverpool's Albert Dock. However, the size of ships using the Port of London rapidly overtook the limited capacity of the lock gates which gave access to Telford's two linked docks, and their commercial viability was relatively short-lived. In the foreground of this photograph is the royal row-barge *Gloriana*, completed in April 2012 for the late Queen Elizabeth II's Diamond Jubilee.

Hazledine also manufactured the chains for Telford's suspension bridges across the Menai Straits and the Conwy estuary – both of which were amongst the earliest suspension road bridges to be constructed anywhere in the world. They carried the main road from Chester to Bangor, part of his much larger project to improve travel between London and Holyhead, and both bridges opened to traffic in 1826.

Telford also designed the castellated tollhouse at the eastern end of the Conwy bridge. The original wrought-iron chains of the Menai bridge were replaced by steel in the years just before the Second World War, but long before that, in 1893, Telford's original wooden roadway deck had been replaced by a steel deck designed by future Forth Bridge engineer Sir Benjamin Baker. The Conwy bridge, however, still hangs from Hazledine's original chains.

While most of Telford's bridges were highly visible, and remain so today, one of his most inventive – across the Bannock Burn at Bannockburn – is, by the nature of the gorge which it spans, all but invisible. Many of those who cross it daily don't even realise it is there. And yet, it is a very special bridge – now a listed structure and more than 200 years old – incorporating an inverted lower arch, the inspiration for which Telford probably drew from mediaeval cathedral builders. That lower arch is, in fact, three arches, used to dissipate the huge loading pressures the gorge would otherwise bring to bear on the main span. Robert Southey described it as 'striking and singular' and it is certainly that.

He was also instrumental in the development of many of Britain's docks, amongst them Ellesmere Port which, when opened, was the largest inland dock in Britain. Although added to and modified in the 200 years since it opened, Telford's layout survives, as do several of the original buildings. The docks are now home to the National Waterways Museum.

Telford's only major project in London was the development of St. Katharine Dock in the late 1820s. The two basins were surrounded by warehouses designed by Philip Hardwick, and the water levels in the basins were originally controlled by two steam engines built by Matthew Boulton and James Watt. Telford's specification for the warehouses required them to

This photograph of the busy Pultenaytown Harbour was taken by Alexander Johnston who had established Wick's leading photographic business in 1863. Shortly thereafter he established a studio in Parliament Square, and this photograph dates from around that time. Telford drew up his early plans for the township around 1807, and the massive double harbour was completed between 1811 and 1812. When the entire project was completed – around 1818 – an estimated 1,000 fishing boats used the port. The building to the left of the photograph was the Temperance Coffee House owned by the Orkney and Shetland Steam Navigation Company which operated steamers to the Northern Isles. The Johnston family operated their studio in Wick for several generations, and their collection of over 50,000 images – many of them the original Victorian glass plates – has been preserved – and is being digitised – by The Wick Society.

be as close to the quay edge as practicable, so that goods could be loaded and unloaded with the minimum of effort. Today Hardwick's warehouses have been redeveloped as apartments, and the basins are home to a mixture of Thames barges and pleasure craft.

Several hundred miles north, he had earlier been in charge of the development of a completely new fishing port and township near Wick in north-east Scotland. The project was the result of the patronage he enjoyed from landowner Sir William Pultenay (1729–1805), one-time Member of Parliament for Cromarty. Pultenay's patronage was also enjoyed by the Robert Adam more than 500 miles south, who had built Pultenay Bridge in Bath for him.

Pultenaytown, as it was named in honour of Sir William who had died in 1805, has rightly been described as one of the earliest custom-designed industry-focused 'new towns' in the world, and it was certainly the first in Britain. The layout of the streets was designed to minimise the impact of the often-ferocious north winds which affect that coast.

The harbour had the capacity to house 1,000 herring smacks, and by the middle of the 19th century had become the busiest herring port in Europe.

But it was not just large-scale projects which Telford undertook. At the entrance gate to the little 19th century parish church on Iona – invariably overlooked by the majority of visitors to the island who are more intent on exploring the world-famous, abbey – is a sign which recalls one of Telford's more unusual projects:

> 'In 1824, the Government engaged Thomas Telford F.R.S. to design and built 32 churches and 43 manses in remote Highland parishes.'

Authorised by Act of Parliament in July 1823, the maximum cost of any one church was limited to £1500, with provision for a £100 stipend to be paid to the minister of each church – a sum later increased to £120. The basic design was for a T-plan church which would enable the congregation to sit as close as possible to the pulpit.

[above] The PS *Gondolier*, built for David Hutchinson in 1866, in one of the clocks at Fort Augustus, c.1905, with Loch Ness and the St. Benedict's Abbey completed in 1880 in the distance. She sailed the Caledonian Canal for the next 73 years in the colours of David MacBrayne. The abbey is now luxury holiday apartments

[top right] A paddle steamer in the flight of locks at the entrance to Loch Linnhe at Banavie. Known as 'Neptune's Staircase – as seen in a postcard from 1906 – it took almost twenty years to construct. The original lock gates were cast by Hazledine at Plas Kynaston in North Wales and shipped by land and water to the canal site.

[above right] Coaches at Muirtown Locks on the Caledonian Canal, photographed in the late 1890s and published as a postcard c.1905. The number of coaches and charabancs awaiting the arrival of a steamer at the quayside at Muirtown near Inverness attests to the popularity of the Caledonian Canal's regular steamer services.

Of those that were actually built, 12 survive – from Portnahaven on Jura to Quarff in Shetland – and most have remained in regular or occasional use. The original plan to build 40 was reduced as only £50,000 pounds in total had been made available by the Government.

The churches were as much a political gesture as they were a social and religious benefit. In the 1830s, the island of Ulva's church, completed in 1828, came in five pounds under budget and could accommodate half the island's population of 570 souls at one sitting. The island's owner in the 1840s claimed that everyone attended regular services – including the island's only catholic, and its single atheist. Only the southern arm is now used as a church, the rest serving as a community hall since the 1950s. The population of Ulva is currently just five, but with the island now community-owned, there are plans to encourage re-population.

The church at Berriedale in Caithness, opened in 1827, was a little cheaper at £1473 but declining congregations over the years led to its closure in 2008. The construction of Iona's church went £3.4s.0d over budget – about £3,500 at today's prices.

Telford's best-known project was probably the Caledonian Canal which cuts diagonally across Scotland, utilising the lochs of the Great Glen and saving mariners the long and often hazardous sail around Scotland's north coast.

He was not the first to suggest such a canal – James Watt had proposed a similar project in the 1770s, while John Rennie had resurrected it 20 years later. But it was a report by Telford in 1801, in consultation with Watt, which had finally received government backing.

The canal was opened in 1822, over 60 miles in length, with 29 locks, designed to accommodate what were considered to be large vessels at the time. Each lock was 170ft (52m) long and 40ft (12m) wide. The author of *Sylvan's Pictorial Handbook to the Scenery of the Caledonian Canal*, published in 1848, noted:

[top left and right] The 'Parliamentary Churches' on the island of Ulva off the west coast of Mull (above left) and at Berriedale in Caithness (above) are just two of the churches designed by Thomas Telford and built with Government money to demonstrate official 'support' for the established Church of Scotland, the supremacy of which had been enshrined in the Act of Union. The church at Berriedale opened in 1827 and cost just £1,473, while the Ulva church cost £1,495.

[left] The design of the dock and warehouse facilities at Ellesmere Port in Cheshire owe much to Telford's work there in the 1830s. He modernised and extended the small port, having previously been involved in the construction of the Ellesmere Canal with William Jessop. While the facilities have been extended and modernised since Telford's days, some of his original buildings still survive.

'The object of the undertaking was to obviate the danger and delay in sailing through the Pentland Firth, and round the north-west coast, by making a direct passage from Fort George to the sound of Mull. It was intended to admit the largest vessels that trade between Liverpool and the Baltic, and even Frigates of 32 guns when fully equipped.'

In the opening years of the 19th century when the canal was being planned, nobody could have foreseen the dramatic increase in the size of ships which would accompany the transition from sail to steam. All too soon, ships outgrew the canal.

A number of other canals suffered from the same problem – the traffic which could navigate them was very quickly constrained by the limited breadth and depth of their locks and cuttings. In the 20th century, however, these canals were extensively restored and have subsequently enjoyed success once more as leisure waterways.

The most interesting book about Telford is his own account of his career, elaborately titled – as was typical of 19th century books – *Life of Thomas Telford, Civil Engineer, written by himself, containing A Descriptive Narrative of his Professional Labours*, which was published posthumously in 1838. He opened his 'descriptive narrative' thus:

> 'Having for more than half a century been constantly employed in planning and conducting works of greater variety and magnitude than fall to the share of most men of my profession, I feel it as a duty incumbent on me to bequeath to posterity a connected description of these operations: for although they have been from time to time recorded in Reports to Parliament, to public bodies, and joint-stock companies, yet this having been done occasionally during a long series of years, it now becomes necessary, in order to render my experiences generally useful, that a careful selection be made from the formidable mass which has unavoidably accumulated...'

There are parts of the text which leave the reader feeling that they have just heard Telford thinking out loud as he considered alternative ways of addressing whatever challenge lay before him. This is very much the case as he describes confronting the engineering challenges back in 1795 of taking the Ellesmere Canal over the River Dee at Pontcysyllte in North Wales – or 'Pont-y-cysylte' as he wrote it:

Herring smacks making their way north towards Inverness and the Beauly Firth, passing through the Tomnahurich Swing Bridge on the Caledonian Canal in 1904. Telford's original 1816 double-sided swing bridge was approaching 100 years old when this photograph was taken, and continued in regular use until a much larger single-ended replacement was installed by Sir William Arrol in 1938 – and that bridge still carries the A82 road across the canal today. The bridge-master's house, now stripped of it harling, still stands by the side of the canal today.

> 'The north bank of the River Dee at this place is abrupt; on the south side the acclivity is more gradual; and here, on account of gravelly earth being readily procured from the adjacent bank, it was found most economical to push forward an earthen embankment, 1500 feet in length from the level of the water-way of the canal, until its perpendicular height became 75 feet; still a distance of 1007 feet intervened before arriving at the north bank, and in the middle of this space the River Dee was 127 feet below the water level of the canal, which was to be carried over it; therefore serious consideration was requisite in what manner to accomplish this passage at any reasonable expense. To lock down each side 50 or 60 feet, by 7 or 8 locks, as originally intended, I perceived was indeed impracticable, as involving serious loss of water on both sides of the valley, whereas there was not more than sufficient to supply the unavoidable lockage and leakage of the summit level. To construct an aqueduct upon the usual principles, by masonry piers and arches 100 feet in height, of sufficient breadth and strength to afford room for a puddled water-way, would have been hazardous, and enormously expensive, necessity obliged me therefore to contrive some safer and more economical mode of proceeding. I had about that time carried the Shrewsbury canal by a cast-iron trough about 16 feet above the level of the ground; and finding this practicable, it occurred to me, as there was hard sandstone adjacent to Pont-y-cysylte, that no very serious difficulty could occur in building a number of square pillars, of sufficient dimensions to support a cast-iron trough, with ribs under it for the canal.'

[top] The north entrance to Telford's Harecastle Tunnel on the Trent and Mersey Canal near Kidsgrove in Staffordshire was cut in order to increase capacity on the busy waterway, and opened to traffic in 1827 after three years' construction. Access to James Brindley's earlier narrow tunnel was via the cut to the right of the picture. Telford's tunnel – like Brindley's before it – is only wide enough for a single canal boat, so traffic is managed by using a convoy system, sending alternate northbound and southbound groups of boats through the 2,900yd (2650m) long tunnel. The roof height diminishes part way through the tunnel, the curved guide hanging at the entrance acting as a warning to boatmen of what lies ahead.

[left] One of Telford's masterpieces, the Engine Arm Aqueduct was built to carry water from Edgbaston Reservoir over his 'New Main Line' canal to Brindley's and Smeaton's parallel 'Old Main Line'. The waterway comprises a cast-iron trough 8ft (2.4m) wide with four feet wide towpaths supported by arcades of gothic-style arches either side.

The cast-iron aqueduct which carried the Shrewsbury canal at Longdon-on-Tern had been completed in 1797, but it was not the first such structure – as with so many innovations throughout Britain's industrial development, others were thinking along similar lines to Telford, and Benjamin Outram's Holmes Aqueduct in Derbyshire was opened just a month before Telford's. Regrettably, it was demolished less than 50 years ago.

Key roles in the creation of the Pontcysyllte Aqueduct were played by both William Hazledine and Telford's mentor, and Chief Engineer on the Ellesmere Canal project, William Jessop. Central to the design was an innovative way of minimising the weight and impact of the bow wave which a boat crossing the aqueduct would otherwise have to push ahead of it. This was a significant added burden, slowing down the horse which was hauling it across. In a narrow channel, quite a substantial weight of water would have built up in front of a heavily laden boat – not perhaps as important an issue in a wider stone-lined canal, but potentially a major problem in the narrow confines of a canal more than 120ft in the air.

[right] A narrowboat crossing the Pontcysyllte Aqueduct creates little disturbance on the water, thanks to Telford's ingenious design of suspending the towpath over the water channel itself.

[far right] The underside of the 'cast-iron trough' which carries the waterway over the Dee Valley. The minimalist supporting structure has now done its job for more than 215 years. Every five years or so, a plug is removed from the trough to allow the water to flow out and down into the River Dee below, in order that essential cleaning and maintenance can take place.

Telford's solution was neat and efficient – he designed the trough to occupy the full width of the aqueduct, which allowed the water displaced by the bows of the boats to move under the towpath, thus dissipating the bow wave. The towpath was originally wooden but was replaced by cast iron plates later in the 19th century.

Thirty years later, John Scott Russell approached the issue of the bow wave from a completely different perspective, and designed the hulls of boats so that they dispersed the water with greater efficiency.

The trough itself is not fixed to the supporting iron ribs, but sits on them, held in place within cast-iron lugs affixed at intervals along either side. It is those lugs, and the weight of the trough itself and the water it contains which keeps it all in place. Describing the aqueduct he would later refer to as his masterpiece, Telford continued

> 'The height of the piers above the low water in the river is 121 feet, their section at the level of high water in the river is 20 feet by 12 feet, at the top 13 feet by 7 feet 6 inches. To 70 feet elevation from the base they are solid, but the upper 50 feet is built hollow; the outer walls being only 2 feet in thickness, with one cross inner wall; this not only places the centre of gravity lower in the pier, and saves masonry, but insures good workmanship, as every side of each stone is exposed. I have ever since that time caused every tall pier under my direction to be thus built.'

All 18 piers were built at the same time – with a few courses of masonry added to each in turn, the scaffolding across the entire span then being raised before the procedure was repeated. Telford believed that such an approach helped the workforce acclimatise to the increasing height, so that by the time they were more than 100 feet above the river, they felt little or no fear.

The individual sections of the cast-iron troughs were fabricated by William Hazledine at Plas Kynaston, Hazledine having built a short length of private canal from Trevor Basin at the north of the aqueduct up to Plas Kynaston to facilitate the movement of ironwork for this and later projects.

This 1806 engraving by Jonathan Parry is inscribed "To Sir Watkin Williams Wynne Bart. This View of PONTCYSYLLTE AQUEDUCT North Wales is respectfully inscribed by his most obedient humble Servt. Jn. Parry." It was also available as a coloured aquatint. The text includes the lengthy inscription on the foundation stone on the central arch, and a detailed description of the bridge itself – "This Aqueduct which extends 988 feet consists of 19 Arches, each 45 feet in the span, with an addition of 10 feet 6in. of Iron Work at each end in continuation; and carrys the water 1000 feet in a cast Iron Trough 11feet 8in. broad. The supporting Piers are Stone, 10 feet wide, and 21 feet deep at the base; and lessening upwards to 7 feet wide and 12 feet deep at the top; their height is 116 feet, which adding 4feet 7in. for the cast Iron Hand Rail and 5feet 6in. for the depth of the trough, makes the total elevation of the Building from the surface of the River Dee 126feet 1in. There are 11 Iron plates screwed together from Centre to Centre of each Arch. And from the springing of the Arch to the bed of the Trough is 8feet 9in.". Some of those measurements differ from those quoted by Telford in his autobiography, see opposite page.(Illustration courtesy of Nigel Phillips Rare Books)

However, according to Samuel Smiles writing in *Industrial Biography,* not all of the project was sub-contracted to Hazledine – some of the ironwork was cast by William Reynolds, son of Richard Reynolds, the former iron-master at of Coalbrookdale:

> 'Telford, the engineer, also gracefully acknowledged the valuable assistance he received from William Reynolds in planning the iron aqueduct by means of which the Ellesmere Canal was carried over the Pont Cysylltau, and in executing the necessary castings for the purpose at the Ketley foundry.'

When completed, the aqueduct cost just over £47,000 which Telford described as 'a moderate sum as compared with what by any mode heretofore in practice, it would have cost'. That was just £5M at today's prices.

The castings for Telford's Longdon-on-Tern Aqueduct, built in 1795, were also made at Ketley in Staffordshire, while the ornate Gothic-style ironwork for the Engine Arm Aqueduct, opened in 1828, was fabricated locally at Horseley Ironworks in Tipton. For such a functional structure, the richness of its decoration and the quality of its execution are of a very high order – none of which would have been visible to the coal barges passing along it. But for those travelling on the New Main Line beneath it, it would have presented a magnificent sight.

Throughout mainland Britain – from Pulteneytown Harbour in Wick to Portsmouth Dockyard on the south coast – there are many hundreds of surviving examples of Telford's prodigious output over a career spanning nearly six decades and embracing roads, canals, bridges, docks, harbours and churches. He has been variously referred to as a 'Man of Iron' and the 'Colossus of Roads', but he was, simply, a master engineer, an innovator who expanded the boundaries of how materials could be exploited. Thomas Telford died in 1834 a month after his 75th birthday, and was buried in Westminster Abbey.

As Robert Southey wrote of his great friend, in his account of their travels together, *Journal of a Tour in Scotland in 1819,* 'In the prime of life, he found his proper place in the world.'

THE REMARKABLE MR. RENNIE

Every day, thousands of people drive across the 200-year-old bridge which carries the A699 over the River Tweed in Kelso in the Scottish Borders without giving it a second glance. Yet this elegant bridge, built between 1800 and 1803, is one of many surviving structures across Britain – and in America where his London Bridge was re-erected in Arizona – designed by the Scottish engineer John Rennie the Elder (1761–1821).

His Waterloo Bridge, opened in 1817 and demolished in the 1930s, was described as 'the noblest bridge in the world' by the Italian neo-classical sculptor Antonio Canova, who also said 'it is worth going to England solely to see Rennie's bridge.' Comparing it with his bridge at Kelso, the Waterloo Bridge can be seen as a more sophisticated evolution of the same design aesthetic. Had it been anywhere other than a busy Thames crossing, this beautiful bridge would have been given the highest level of historic buildings protection rather than being demolished – reportedly due to subsidence under one of the piers – and replaced with a wider carriageway.

While the names of both James Watt and Thomas Telford are familiar to most people, John Rennie's name has invariably been in the background, yet his impact on the transport infrastructure of Britain is equally important.

[above] A portrait of Rennie from c.1811 when he started work on London's Waterloo Bridge.

[below left] Rennie's bridge over the Tweed at Kelso.

[bottom] Waterloo Bridge, opened in 1817, was described as 'the noblest bridge in the world' by the sculptor Antonio Canova.

[opposite page] A busy scene at the top of John Rennie's remarkable flight of locks at Caen Hill on the Kennet & Avon Canal in Wiltshire.

[inset A view from the foot of the Caen Hill flight looking up towards Devizes. Such was its demand for water at the canal's busiest times, that each lock on the flight was built with an adjacent lagoon to the left which both fed the lock and accommodated waiting narrowboats.

connections with:

- James Watt
- John Smeaton
- James Brindley
- William Jessop
- Thomas Telford
- William Fairbairn

Preston Mill just outside the village of East Linton in East Lothian – was maintained by Meikle for several years, perhaps even assisted by the young Rennie. It has been beautifully preserved by the National Trust for Scotland.

Rennie was the youngest child of James – a brewer and tenant farmer – and Jean Rennie, and was born at Phantassie Farm near East Linton in East Lothian in June 1761. John was only five years of age when his father died, and he was raised by his mother and his elder brother, George.

From an early age the boy showed a fascination for anything mechanical – and almost on his doorstep were several sites which were very mechanical – Preston Mill and Houston Mill on the River North Tyne. Rennie's farmer brother would probably have been a customer of one – or both – of them.

In charge of the Houston Mill – now a private house – during Rennie's childhood was another eminent Scottish engineer, the millwright Andrew Meikle, who is best known for inventing (or perhaps just improving) the threshing machine around 1786. Meikle had, in the early 1770s, also designed a revolutionary system of slatted sails for windmills, hugely improving their efficiency, and allowing them to be adjusted mechanically to maximise efficiency in changing wind conditions.

Houston Mill, however, like Preston Mill nearby, was water-powered. Preston Mill – which was maintained by Meikle for several years, perhaps even assisted by Rennie – has been beautifully preserved by the National Trust for Scotland and is regularly open to the public.

Meikle took the young Rennie under his wing in 1773 teaching him the basics of mechanics, especially with regard to mills. A millwright in those days was usually someone involved with the construction, operation and maintenance of corn mills, but as the 18th century drew to a close and industrialisation accelerated, millwrights were in demand in much larger industrial mills.

It seemed like an ideal career, and at the age of just 18, John Rennie set himself up in business as a millwright, managing the demands of his growing enterprise while pursuing academic studies at Edinburgh University.

After graduating with a degree in Natural Philosophy and Practical Sciences, in 1783 Rennie embarked on a tour of some of the industrial centres of England. One of the places he visited was Boulton and Watt's Soho Foundry which had been established eight years earlier. There he met James Watt who was developing a large condensing steam engine to be installed in London's Albion Mill, the world's first steam-powered corn mill being constructed at the time by Boulton. Watt must have been impressed by what he saw in Rennie, for the following year he offered him a job.

At that time, Boulton and Watt were assembling condensing steam engines of their own design, but not yet manufacturing all the components themselves. They would not bring the whole manufacturing process under their own roof until 1795. Rennie's role seems to have been supervising the installation of the first of these steam giants, and one of the first projects with which he was involved was the erection and equipping of Albion Mill.

Despite this early commitment to mill machinery, Rennie's career as a millwright was short-lived, and by 1790 he had been appointed Surveyor to the Kennet & Avon Canal Company whose waterway was planned to run from Bristol to Reading. This was an early example of his many achievements as a canal and bridge builder.

To build the canal, Rennie had to overcome major challenges – bridging gorges, excavating tunnels and climbing hills – as well as harnessing sufficient water to keep the canal operational. He had to cross the River Avon twice – for which he built the Dundas and Avoncliff aqueducts – and build two pumping stations to raise water up to the higher reaches of the canal to keep water levels topped up.

The Claverton pumping station was a truly ingenious solution to a problem imposed on the canal-builder by the geology of the landscape through which the canal was being driven. The canal bed leaked persistently. Rennie's usual solution – puddling clay into the base and walls of the canal – did little to stem the leaks, so rather than face the challenge of regularly closing and draining sections of the waterway to apply yet more clay, he offered the pumping station as a radical alternative. Using nothing more than the power of the River Avon itself, the station pumped water from the river 48 feet up to the canal. For the pumps themselves Rennie installed two large water-powered beam engines.

At the time the canal was being built, there was already a grist mill fed by a lade from the River Avon, so the potential for a water-powered pump was evident. Here was the ideal opportunity for Rennie to call upon his earlier experience as a millwright, and design the pumping station himself. His design was for a breast-shot watermill powering two large beam pumps – a very elegant solution to the challenge – and, unlike Crofton, it was cheap to run having zero fuel costs. Rennie designed the mechanism himself, using Boulton and Watt beams.

Part of his redesign of some of the canal's route was the replacement of the planned horse-drawn railway incline up Caen Hill near Devizes with the magnificent flight of 16 locks which raise the canal up the steep hill. The Caen Hill lock system is actually three groups of locks raising the canal 237ft (72m) in just under two miles (3.2Km).

The flight, which was completed in 1810, required the creation of large lagoons to one side of 15 of the locks, both to store the huge amounts of water needed to operate them, and to provide 'parking' for boats waiting to enter the next lock up or down.

Further along the waterway, at Crofton, a steam-powered pumping station was installed to feed water to one of the highest stretches of the canal and, given his long association with James Watt, it is no surprise that the first steam engine to power it was ordered from Boulton & Watt.

Crofton, which was operational by 1812, is unique – the last surviving steam-powered pumping station still with its original beam engines intact, and still capable of doing the job for which the station was built. Of course, for most of the time these days pumping water into the highest reaches of the Kennet & Avon Canal is done by electric pumps, but during the several weekends each year when the station is in steam and open to visitors, the electric pumps are shut down, and the early 19th century engines resume the task for which they were installed.

The Kennet & Avon was not Rennie's first canal project. In 1801, his Crinan Canal was opened, construction work having started on it in 1794. Three years before that, in 1791, he had surveyed the route of the Rochdale Canal in Lancashire – his first major involvement in canal design – and history has attributed to him the roles of both surveyor and engineer, unfortunately diminishing the major involvement of William Jessop.

Construction work on the Rochdale Canal and the Crinan Canal started within months of each other – building the 32-mile Rochdale Canal taking ten years, compared with seven years for the nine miles between Ardrishaig and Crinan.

But Rennie's involvement with plans to make a cut across the Kintyre peninsula had started long before that. As a 20-year-old in 1781 he had been part of a team involved in surveying possible routes to avoid the long sail around Kintyre.

Just over 200 years ago, any captain leaving the port of Glasgow bound for Oban or the Western

Matthew Boulton's Albion Mill, as illustrated in the *New London Magazine* in 1786. The Southwark mill was designed by Matthew Wyatt and engineered by John Rennie, with engines built by James Watt. It was the large first steam-driven flour mill ever built and could produce as much flour in a month as its biggest rival could produce in a year, That massive output was not welcomed by other mill owners who feared being put out of business. It lasted just five years. On 2 March 1791 it was completely gutted by fire. Arson was suspected.

The ALBION MILL, *Blackfriars Bridge.*

[opposite page] The Dundas Aqueduct on the Kennet & Avon Canal, located on the Somerset/Wiltshire border a few miles from Bath, is one of two beautiful aqueducts Rennie built using Bath stone to carry his canal over the River Avon. Work on the Dundas Aqueduct started in 1795, the Avoncliff in 1797; both were complete by 1801. When built, both aqueducts – the Avoncliff is about two miles east – were constructed with three arches. But then along came Isambard Kingdom Brunel who wanted his Great Western Railway to run beneath the canal as well – the tracks can be seen beyond the river in the lower picture – resulting in a fourth archway being cut through to the north-west of the river in the 1840s. In 1951 the Dundas Aqueduct became the first canal structure in Britain to be awarded Grade 1 listed status.

[above] The Avoncliff Aqueduct is a similar but slightly less ornate structure than the Dundas Aqueduct, opposite.

Isles, had a long and sometimes daunting voyage ahead of him. Down the Clyde, past the islands of Bute and Arran, down the east side of the Mull of Kintyre, round the often treacherous seas at the tip, and then the long sail up Scotland's west coast. Laden with coal, and an assortment of other general supplies, the little vessels which plied their trade between the Clyde and Scotland's islands were often a poor match for the seas and the winds which hampered their passage.

On the return journey, bringing fish and an assortment of island produce back to Glasgow, the major problem was one of time – the journey was, simply, too long and too slow to be really profitable.

For the fisherman bound for the fishing grounds in the northern sector of the North Sea, the voyage from the Clyde not only involved that journey around Kintyre, but the negotiation of the seas around Cape Wrath at the north west tip of Scotland, and the treacherous swells of the Pentland Firth.

Towards the end of the 18th century getting to the Pentland Firth was not the challenge occupying the minds of a group of local businessmen, headed by the Duke of Argyll. Their idea was simply – although it turned out to be anything but simple – to build a canal across the narrow Mull of Kintyre. Their primary concern was to reduce sailing time between Glasgow, Oban and the Western Isles.

Two routes were considered – the shorter one across the narrow strip of land which separates East Loch Tarbert from West Loch Tarbert, and the longer running from Ardrishaig to Port Righ at the top of Kintyre. Although requiring a much longer canal, the second route was favoured as it reduced the sail around Kintyre by much more than the shorter, but more southerly, route. When built, the nine mile long canal saved over 130 miles of difficult sailing, and greatly reduce the journey time between Glasgow and the islands. Even after all the necessary approvals had been granted and Acts of Parliament processed into law, it took four more years for the Parliamentary Commissioners to approve the formation of a company to develop the project, and a further 16 years before the canal opened.

Once the northern route had been selected, two options were considered, the first being from Ardrishaig to Port Righ (now known as Crinan) and the other longer route to Duntrune on the north side of Crinan Loch. Had the longer route to Duntrune been chosen, the canal

[top] Claverton Pumping Station and its weir, sitting almost 50 feet below the canal it was designed to feed.

[middle] The massive water wheel which drives the mill is actually two linked wheels side by side.

[bottom] The Boulton and Watt beams which drive the pumps at Claverton.

would have had to traverse Moine Mhor, the unstable mossy landscape on either side of the River Add. Work started in 1793, and the planned duration of the project was a mere three years, but that proved wildly optimistic.

Even the chosen route, along the edge of the moss, was fraught with construction difficulties. Indeed, the whole building project took more than twice as long as expected, with hard rocks to be cut through for the descent down into Crinan, and soft peaty bogs to be stabilised between Dunardry and Islandadd Bridge.

As the canal was built before the age of steam, the towpath had to withstand the constant pounding of horses' hooves. By the time the canal eventually opened to traffic in 1801, it was still incomplete, and the presence of increasing amount of coastal shipping in those first years of operation only served to exacerbate the problems. The rubble masonry construction of the canal sides and towpaths collapsed in several locations, the American oak timbers of the lock gates rotted prematurely and had to be replaced. By 1805 a completely new stretch of canal had to be constructed between Cairnbaan and Oakfield so severe had been the erosion of the original section.

It was 1809 before the project was deemed 'complete', but it was not Rennie who was charged with sorting the problems out. That job went to Thomas Telford, who was already also engaged in what would be his largest project in Scotland – cutting waterways to link the lochs in the Great Glen as part of the project which would become the Caledonian Canal. While never sustaining the hoped-for level of commercial success, both canals have been in constant use, albeit today used mainly by pleasure craft.

Queen Victoria was impressed when she and Prince Albert used the Crinan Canal to traverse the Mull of Kintyre on her Royal Barge in 1847. She found the concept of locks fascinating but tiresome, especially as the planned 90-minute passage – pulled by four scarlet-liveried horses – took nearly twice that long. Arriving at Loch Crinan after eight o'clock in the evening, she made straight for her royal yacht, the *Victoria and Albert*, and bed.

Amongst Rennie's many other projects was the development of London Docks, as well as three important bridges across the Thames – at Waterloo and Southwark and, of course, London Bridge.

A boat making its way through the five lock system at Dunardry on Rennie's Crinan Canal. The Dunardry Locks mark the beginning of the canal's descent down to Loch Crinan. The first fishing boat passed through the canal in 1801, and after Queen Victoria's visit, the canal became part of David MacBrayne's 'Royal Route to the Isles', with more than 40,000 passengers a year sailing along it on the small steamship *Linnet*.

[right] London Bridge under construction in November 1827 with the Lord Mayor's Procession passing under the incomplete arches at a very low tide level. The illustration was drawn by Thomas Hosmer Shepherd (1793–1864) and engraved by Thomas Higham (1795–1844). Rennie himself had died six years earlier.

[below] The huge beam on the 1846 engine by Harvey & Company of Hayle, at Crofton Pumping Station.

[below right] The Cylinder Head Room at Crofton Pumping Station.

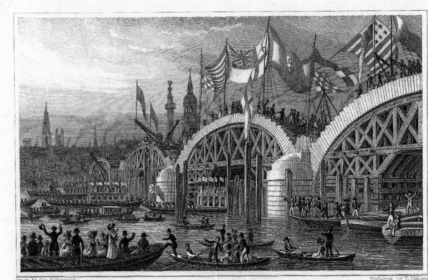

NEW LONDON BRIDGE, WITH THE LORD MAYOR'S PROCESSION PASSING UNDER THE UNFINISH'D ARCHES, NOV^R 9, 1827.

Published Aug. 16, 1828 by Jones & Co 3, Acton Place, Kingsland Road, London.

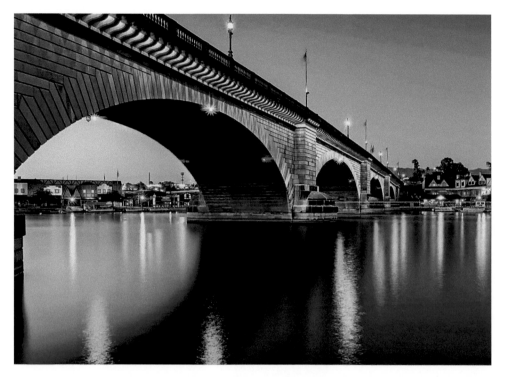

[left] Although designed by Rennie, London Bridge was not completed until 1831, ten years after his death. It was completed under the supervision of his son, Sir John Rennie, also an eminent engineer. In 1968, due for demolition after signs of subsidence appeared, it was sold to American oil millionaire Robert P. McCullough. The facing stones were dismantled, numbered, and shipped to America, and 'London Bridge' now stands across the Colorado River in the city McCullough created – Lake Havasu City, Arizona. *Photo: Juan O'Neal/Go Lake Havasu Tourism Bureau.*

[below] The 1882-built former HMS *Dolphin* moored in Rennie's West Old Dock at Leith. The dock has since been infilled, and in order to salvage the copper cladding which had protected her hull since launch, *Dolphin* was towed to Bo'ness in July 1977 and unceremoniously set alight. Both Rennie's East Dock and West Dock measured over 750ft (228m) in length and around 300ft (91m) wide. At the time of their construction, those were large enough to accommodate most ships.

He was also involved in the redevelopment of Portpatrick Harbour, Greenock's East India Docks and works on the Broomielaw in Glasgow, and a major expansion of Leith Docks – where, in addition to his continuing work in England, he was Project Engineer on the new East Dock from 1800 to 1807 and again on the West Dock between 1810 and 1817.

It was the creation of these docks which turned Leith into a truly commercial port, allowing ships to load and unload their cargo at the quayside for the first time, and at all stages of the tide. Hitherto, it had been necessary for ships to stand in the estuary while their cargo was brought to shore in small boats.

Rennie's docks were the first 'floating harbours' or 'wet docks' to be developed in Scotland and they triggered a huge expansion of trade to and from Edinburgh's port in the 19th century.

Subsequent developments at Leith in the 20th century resulted in both his docks – by then known as the East and West Old Docks – being infilled in the 1970s. All that survives are the remains of the entrance locks. The site which the docks once occupied is now home to government offices.

John Rennie died in 1821 at the age of just 60 and, despite all his achievements – he was just as celebrated in his day as were Jessop and Telford – over time he became an almost-forgotten member of the pantheon of great Scottish engineers – until 2014, that is, when he was at last inducted into the Scottish Engineering Hall of Fame.

On the quayside in Bristol's Floating Harbour, the 35-ton tubular steam crane is the last working example of Fairbairn's patented design – although several manually-cranked examples also survive. Now a listed structure, it was built by Stothert & Pitt of Bath, and is still occasionally steamed.

WILLIAM FAIRBAIRN – TUBULAR BRIDGES AND LANCASHIRE BOILERS

On the quayside at Wapping Railway Wharf, in the heart of William Jessop's Floating Harbour in Bristol, stands a unique crane, built in 1878 by the important Bath-based crane builders Stothert & Pitt. It is the last operational example of a steam-powered tubular crane surviving in Britain. With a lift capacity of thirty-five tons, and the ability to be rotated through a full 360 degrees – allowing it to be rotated directly over the cargo holds of the largest ships of the day and then discharge the cargo into wagons on the dockside railway – it was one of the most versatile cranes in the world. Patented in 1850 by the eminent engineer William Fairbairn (1789–1874), it is of a design which could once be seen in use in many countries. In his specification (British patent No.13,317) Fairbairn described it thus

> 'My Improvements in Cranes and other Lifting or Hoisting Machines consists in constructing the jib of such machines of metal plates so arranged and combined as to form a connected series of tubular or cellular compartments instead of forming the jib of one solid piece as heretofore.'

The design combined light weight with enhanced rigidity, and offered improved operational versatility and much greater lifting capacity than contemporary alternatives. It was also an elegant piece of engineering, with the chains and gearing housed within the tubular structure and thus protected from the effects of weather.

[below left] The Lancashire boilers at the Lady Victoria Colliery in Newtongrange, – the fuel burned in the furnaces was the slurry which came from the on-site coal washeries – the wet dross which could never be sold. Effectively, therefore the colliery machinery was powered free of charge. There was a cost, of course, in the staff who had to fuel these voracious boilers round the clock, shovelling the muddy mixture which poured into the fuel hoppers behind them. A hot, damp, back-breaking job, working in conditions as unpleasant as those underground.

[below right] Sir William Fairbairn, First Baronet Ardwick.

connections with:
- John Rennie
- James Watt
- William Jessop
- Thomas Telford
- The Railway Stephensons
- Isambard Kingdom Brunel
- John Scott Russell

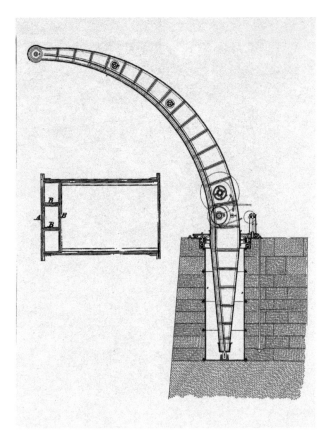

[above] A profile diagram of Fairbairn's patented crane design, showing the 35˚ radius of the 40ft high (12m) tubular jib, and the 25ft (7.5m) tapered section which sat in a deep well in the quayside. The inset shows the four-box construction of each segment, the three small boxes adding strength to the concave face, constructed from riveted wrought iron. Early models were hand-cranked, but the heavy-lift versions were fitted with a two-cylinder steam engine

[above right] W. Fairbairn and Sons exhibited a range of their machinery at the Great Exhibition in the Crystal Palace in 1851. In this contemporary aquatint of the machinery hall, their corn mill, and the end of the jib of the tubular crane can be seen.

To give his 'heavy-lifting' version of the crane greater resistance against flexing as it took the strain, Fairbairn specified that the concave face of the jib should be strengthened by using a girder, the internal structure of each which contained three small and one large box. This was an evolution of his concept for tubular bridge construction which he had outlined in an illustrated letter to George Stephenson in 1845. Perhaps surprisingly, and despite clear documented evidence of Fairbairn's involvement, popular history has attributed the tubular bridge to Stephenson.

An early hand-cranked example of the tubular dockside crane was exhibited on the 'William Fairbairn and Sons' stand at the Crystal Palace Great Exhibition in 1851 and, in his subsequent report on the exhibition to the British Association, Sir David Brewster observed

'These structures indicate some additional examples of the extension of the tubular system, and the many advantages which may yet be derived from a judicious combination of wrought-iron plates, and a careful distribution of the material in all those constructions which require security, rigidity, and strength.'

Kelso-born Fairbairn's pedigree was impressive – aged just 14 he had worked for a few weeks as a mason's labourer on the construction of John Rennie's Kelso Bridge in 1803, and had become acquainted with George Stephenson in 1805. The two men subsequently developed an enduring professional relationship.

Aged 21, he gained employment with Rennie in London but was precluded from taking up his post because of the closed shop policies of the powerful mechanics unions. Returning to London the following year, however, he did find work – and sound mechanical training – at Penn & Son's Foundry in Greenwich.

At the Great Exhibition, Fairbairn exhibited his innovative design for a corn mill, his patent steam-powered riveting machine, and two of the company's railway locomotives – one for hauling freight and another for passenger service. One of the exhibits was the company's 2-2-2 tank engine (*illustrated below*).

W. Fairbairn & Sons had entered the railway market in 1839, their first locomotives being supplied to the Manchester & Leeds Railway, for whom they would build 69 passenger and goods engines, often variants on established designs by Bury, Curtis & Kennedy, Sharp Bothers (later Sharp, Stewart & Company), and others.

In total they built fewer than 400 locomotives between 1839 and 1863 when the railway division was sold to Sharp, Stewart & Co.

Fairbairn made significant contributions in other areas of industrial development – he built mills, bridges, steam engines and, of course, is credited with significant modifications to the design of the steam boiler, resulting in the ubiquitous Lancashire Boiler.

Like James Watt's steam engine, the hugely successful Lancashire Boiler was created by bringing together a number of separate innovations to create something

WATER STREET RAILWAY BRIDGE, MANCHESTER. BUILT 1829.

which was greater than the sum of its parts. The Lancashire Boiler offered control and consistency in the delivery of steam power of an order far in excess of what had been possible with earlier designs. But rather than being a sudden 'Eureka-moment' in boiler history, its design evolved as a result of decades of innovation and improvement by dozens of different engineers. It was the subject of many patents over its long service life, and thus getting a clear picture of its evolution is difficult, with several different 'inventors' and timelines to be unravelled.

However, the pre-eminent 'catalysts' were Fairbairn and John Hetherington who patented aspects of the design and operation of the boiler under British Patent No.10,166 on 30 April

[top] From one of F. Moore's series of railway postcards, one of Fairbairn's early 2-4-0 locomotives, as later modified with a cab. No examples survive in Britain, but some of his locomtives designed in the 1850s survive in Portugal and Brazil.

[above] Water Street Bridge in Manchester, on the approach to Liverpool Road Station, was built for the Liverpool and Manchester Railway in 1829. It was unusual, being built of flat cast-iron beams on cast-iron columns – cast in Fairbairn's foundry – rather than the more conventional arches. It was demolished and replaced in 1905, and this postcard was published in January that year.

[left] Fairbairn's 2-2-2 tank locomotive exhibited at the Great Exhibition in the Crystal Palace in 1851, was illustrated as Plate 89 in Volume 4 of *The Practical Mechanic's Journal*.

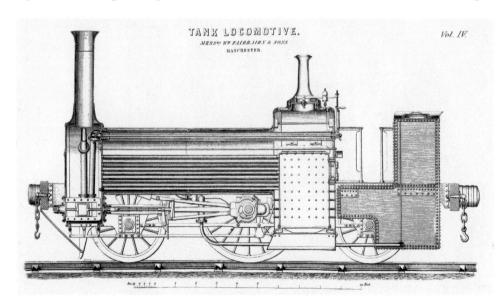

TANK LOCOMOTIVE.
MESSRS WM FAIRBAIRN & SONS
MANCHESTER.

Vol. IV.

[top left] The waterwheels at Catrine Mills in Ayrshire were known as 'The Lions of Catrine'. In the 1878 edition of his book *A Treatise on Mills and Millwork*, Fairbairn described the wheels – believed to be the largest waterwheels in the world. 'The total supply of water requisite to work the mills when the wheels were started was about 60 tons, or 2,150 cubic feet per minute, the wheels revolving at a circumferential velocity of 4 feet a second, or 182 buckets passing each sluice per minute. When working to their full power of 240 horses... ...this pair of wheels would require 98.2 tons of water per minute... The wheels are 50 feet in diameter, 10 feet 6 inches wide inside the bucket and 15 inches deep on the shroud; the buckets are 120 in number.' They were used until the 1940s, although augmented by steam engines to meet the power demands. The first – a 30hp (22kW) Boulton and Watt sun and planet engine – had been ordered by David Dale in 1800, 25 years before the waterwheels were installed.

[top right] Fairbairn's first patent – British Patent No.7,302 for his steam-powered riveting machine, patented on his behalf by Robert Smith in August 1837 – was the first of nine between 1837 and 1873.

[right] The Lancashire boiler at Crofton Pumping Station in Wiltshire was built at the GWR's Swindon Works in 1899 and installed at Crofton in 1987. It provides steam for the station's 1812 Boulton & Watt beam engine and its 1846-built partner by Harvey of Hayle.

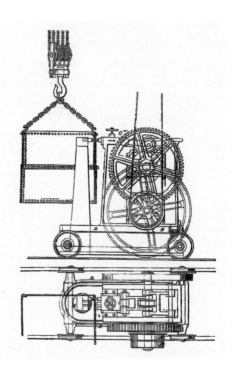

1844, titled '*Certain Improvements in Stationary Steam Boilers, and in the Furnaces and Flues connected therewith*'.

Early boilers delivered steam at low pressure – James Watt never saw high-pressure steam as the future – and as late as 1844, the boiler installed in Brunel's iron steamship, the SS *Great Britain* only delivered steam at a few pounds per square inch. And yet, that is the same year that Fairbairn and Hetherington patented their 'improvements' to what became known as the Lancashire Boiler, its name chosen to distinguish it from the 'Cornish' boiler which could trace its origins back to Trevithick and Newcomen.

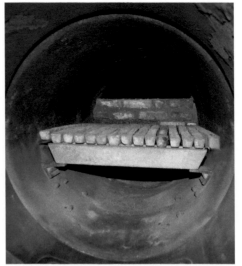

[far left] On the Meldrum Brothers' Patent Furnace, installed at Hereford Waterworks in 1894, air intake was controlled by adjustable dampers located below the fireboxes rather than in the fire door fronts.

[left] Inside one of the two fire tubes on the Lancashire boiler at John Rennie's Crofton Pumping Station on the Kennet and Avon Canal in Wiltshire.

The Lancashire boiler could be operated at a much higher pressure, and once the manufacturing challenges of improving its safety and reliability had been ironed out, the commercial potential of the steam engine was assured.

Fairbairn and Hetherington would, however, have been unlikely to be granted their patent under today's regimen, as a thorough search of contemporary practices would have challenged their claim that

> '[We] do hereby declare the nature of our Invention to consist in constructing stationary or land steam boilers, with two large open tubes passing through each of them below the level of the water to be contained therein, which tubes are parallel to each other, or nearly so, and unite the front part to the back part of each boiler'.

[below] One of the Lancashire boilers at Coldharbour Mill in Devon. This one was built in 1910 by Galloways of Manchester who once even threatened William with legal action for breach of *their* patents.

Multi-tubed boilers had, in fact, been introduced 16 years earlier by Robert Stephenson & Company who had fitted one large and two small flue tubes into their 1828 locomotive *Lancashire Witch*.

Significant in Fairbairn and Hetherington's patent, however was the fuelling sequence, requiring the furnaces to be stoked alternately, with one fire door being closed before the other one was opened. This not only improved combustion, but also reduced both smoke output and soot build-up in the flues.

To get even higher pressure, the techniques of boiler manufacture had to progress considerably. Early boilers would, simply, have exploded if the steam pressure became too high, the methods of their construction giving them inherent weaknesses. Indeed, William Pole in his 1877 biography of Fairbairn wrote

> 'The great extension of manufacturing industry in the Lancashire and Yorkshire towns had led to the employment of steam-power to a vast extent; steam-engines were required in great numbers, and their manufacture was often undertaken by persons not well instructed in scientific principles, and at prices which did not admit of all possible care being taken in regard to the proportions

[above] Fairbairn and Stephenson's tubular bridge which carried the Chester and Holyhead Railway across the river at Conwy, alongside Telford's suspension bridge. It is the only surviving example of the tubular bridge construction they pioneered, the tubular structure of the Menai Straits bridge having been destroyed by fire in the 1970s. Fairbairn was appointed Superintendent in charge of construction on both bridges. Each of the Conwy tubes weighed 1,300 tons and they were floated on pontoons into position beneath the bridge before being raised by hydraulic rams.

[above right] Fairbairn engineered the iron and brick 'fireproof' interior construction of the huge Saltaire Mill complex near Bradford, built for Sir Titus Salt.

or the practical workmanship. Moreover, these engines were not unfrequently worked under careless management, being put into the charge of incompetent or ignorant men, unable to see where danger arose, or unscrupulous as to overtaxing the powers of the apparatus. Hence, boiler explosions became but too common in these districts; and when they did occur, from the magnitude of the buildings and the great number of people employed, the consequences were usually very severe. The worst feature of the case was that the causes of these explosions were often very difficult to trace out. The destruction was so complete that tangible evidence was in a great measure destroyed; and it usually happened that the persons who would have been best able to throw light on the causes were killed.'

Despite those problems, the Lancashire boiler became the boiler of choice, especially in mills and factories across the north of England. By comparison, the Cornish boiler was a simpler device – a large tank for the water, with a large fire tube running part way through it, heating the water to raise the steam. It had a number of inherent issues, especially the shallow distance between the top of the fire tube and the top of the water jacket. If the water level dropped sufficiently to expose the top of the fire tube, an explosion was almost inevitable.

The Lancashire boiler, with its two smaller fire tubes side by side instead of a single large one, had more water above the tubes so less likelihood of disaster if the water level dropped a little. It also raised steam more quickly.

There was a larger grate area which speeded up the raising of steam, and by setting the fire tubes lower within the water tank, there was a greater area for heat transfer from the heated flue tubes to the water. That in turn also aided circulation of the water itself.

Perhaps Fairbairn's greatest contribution was in analysing how the heat transfer from furnace to water actually took place, as a result of which he proposed re-engineering the boiler in such a way as to maximise the efficiency of that process.

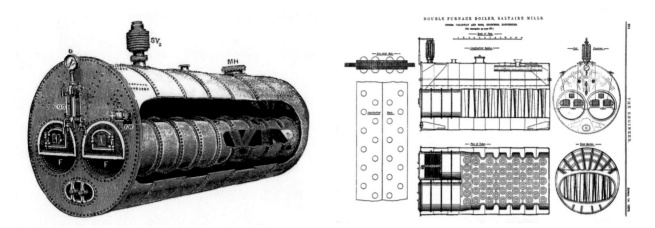

His proposals included routing the exhaust gases from the furnaces underneath the water tank – effectively using the heat twice – and then circulating those same gases around an outer jacket – a third use – before venting them out through the flue.

Another significant improvement – and patented by him although not an original idea at all – introduced air into the system to improve combustion and importantly in the most heavily-industrialised towns, that draught of fresh air considerably reduced the smoke and exhaust gases being vented into the atmosphere.

That simple idea had actually been pioneered a few years earlier by John Wakefield, a Manchester engineer, and by Charles Williams who also suggested that the firing and stoking of the twin fire tubes should be done sequentially rather than in tandem. Their experiments had proved that this significantly aided the digestion of smoke. Research into the role air plays in the combustion process and into training stokers how to fire a boiler so that the right amount of coal was added at just the right time, not only improved the efficiency of the boilers by several percentage points, but also made possible large saving in the amount of fuel burned.

[top left] The Lancashire boiler with its twin fire-tubes set just below the boiler's centre-point, created space for a much larger water jacket and more consistent steam output than single-furnace 'Cornish' designs. This sectioned view of a Galloway's Patent Lancashire Boiler comes from the 1890 edition of *An Elementary Manual on Steam and the Steam Engine* by Andrew Jamieson, published by Griffin & Co.

[above] The Galloway boiler's internal angled tubes, increased the surface area between the flue gases and the water, speeding up steam generation.

[left] Awaiting restoration, this huge Lancashire boiler at the Lady Victoria Colliery (now the Scottish Mining Museum) has had some of its top plates removed, revealing the twin fire tubes within. The boilers were acquired secondhand in 1924, having previously produced steam at a cordite factory near Gretna. Built by Tinker, Shenton & Co. of Hyde and operated at 160psi, they were fitted with Green's economisers which pre-heated the water before it was fed into the boilers, thus improving efficiency.

[above] Two Swiss entrepreneurs, Franz Carl Caspar and Johann Jakob Lämmlin, ordered PS *Minerva* in 1834. According to Charles Frederick Young's account, after construction, she was dismantled and transported to Hull – other sources suggest it was to Selby – where she was rebuilt and sailed via Rotterdam up to Rhine Falls, dismantled again, shipped by road to Lake Zurich, and re-assembled at the lakeside. She entered service in 1835, the first iron steamer on any Swiss waterway. Aquatint by Jakob Hausheer (1813–1841).

[above right] Originally built as a frigate, the 2,057 ton HMS *Magaera*, launched in 1849, never saw service as a fighting ship due to the fragility of her hull plating. Instead her armaments were removed and she was was converted into a supply ship, saw service in the Crimean War, and was broken up in 1867. She was one of the first iron-hulled ships built for the Royal Navy, and one of the last built by Fairbairn at the Millwall yard. The painting on which this lithograph is based was created by one of her crew, Midshipman Thomas C.D. Thompson, and the original is in the collection of the National Maritime Museum.

But the Lancashire boiler was by no means Fairbairn's only game-changing innovation. If he believed there was a way of improving something – as with the boiler and the tubular cranes and bridges – he embarked on endless experiments in search of a better solution.

The pioneering work on the use of iron in shipbuilding, carried out both in Manchester and at his Millwall shipyard which he opened in 1836, is said to have inspired Brunel when he was planning to build the SS *Great Britain*.

Amongst other things, his experiments explored the longer-term effects of flexing on iron plates caused by the rolling motion of a ship at sea. As ships got ever-larger, such considerations became more and more important.

Fairbairn's Millwall Iron Works was the first yard on the Thames to build iron ships, and the screw steamship *Sirius*, built in 1838 to operate on the River Rhone, became the first iron-hulled vessel to be launched on the river – and the first iron-hulled vessel to be listed on Lloyd's Register.

Fairbairn's *Sirius* should not be confused with the more famous wooden-hulled paddle steamer of the same name launched the previous year by Robert Menzies & Sons of Leith which became the first ship to cross the Atlantic entirely under steam. Originally intended for Irish Sea crossings, the 703-ton paddle steamer, chartered by the British & American Steam Navigation Company, sailed from London to New York by way of Cork in 1838 with 40 passengers. When she ran out of fuel approaching Sandy Hook, New Jersey, her captain refused to hoist sails and, instead, burned some of her wooden spars. She beat Brunel's *Great Western* to New York by a few hours.

Several years before he opened the Millwall yard, Fairbairn had started building iron boats in Manchester. In his book *The Fouling and Corrosion of Iron Ships: Their Causes and Means of Prevention*, published by the London Drawing Association in 1867, Charles Frederick Young wrote

'In the year 1830, Dr. William Fairbairn, in conjunction with his partner Mr. Lillie, built at Manchester the iron paddle-boat "Lord Dundas," for use on the Forth and Clyde Canal. The hull was 68'x 11' 6" × 4' 6", built of light angle and T iron, with plates 1/10 thick, and had engines of 10 nominal horse-power, working a paddle-wheel placed in the centre of the boat, in a wheel trough extending the whole length of the hull. The light draught of this boat was 16", and she could accommodate 100 to 150 passengers. Dr. Fairbairn informs the author that he made the voyage from Liverpool to Glasgow in her; and she proved so successful that his firm built eight more of a larger size within the next two or three years for Scotch canals; two passenger-boats with 40 nominal horse-power engines for the Humber; and two for the lakes of Zurich and Wallenstadt in Switzerland, which, after being tried, were taken to pieces and sent out...

... The difficulties which were found to exist in an inland town like Manchester for the construction of iron vessels led to the removal of this branch of the business to London in the years 1834–5, where, at the works, Millwall, Poplar, Dr. Fairbairn constructed upwards of eighty vessels of various sizes, including the "Pottinger," of 1250 tons and 450 nominal horse-power, for the Peninsular and Oriental Company; the "Megæra" and other vessels for the British Government, and many others; thus introducing iron shipbuilding on the river Thames; and in 1848 he retired from this branch of his business.'

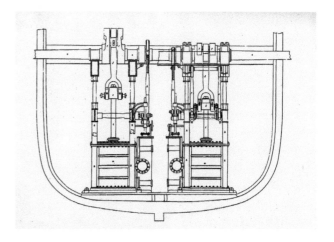

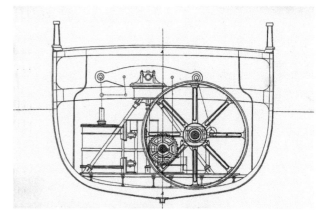

Amongst his innovations in iron shipbuilding was the introduction of transverse watertight bulkheads which had the dual role of strengthening the hull, and protecting the vessel from catastrophe should any puncture of the hull happen below the waterline. From early on in his shipbuilding career, Fairbairn used high pressure steam engines, largely of his own design – aspects of which he protected with two patents – No.9072 of 1841 and No.10,095 of 1844.

While his engineering prowess was considerable, the shipbuilding business was not a commercial success. Operating a shipyard in London while the rest of his interests were in Manchester was clearly fraught with difficulties, and no doubt contributed to the decision to sell the yard to John Scott Russell who with Brunel, would later build the SS *Great Eastern*.

Among his research interests, was the study of the implications of using steam-power on canal barges, and the challenges posed by their bow waves. That was an area of study which Russell – and after him William Froude (1810–1879) – would help develop into more efficient hull designs.

The range and importance of his engineering achievements is immense, but he was a self-effacing character, one of whose obituaries summed up his life with the recognition that:

'Affable and accessible, he was ever ready to communicate information and to give advice to all who sought it, buoyant and cheerful, he had the happiness to attract the esteem and affection of all with whom he came in contact...

... He was, however, singularly modest and unassuming, and used to say, with characteristic self-deprecation, that "any man might do all that he had done, and more, if he would only study and work."'

[Top] Fairbairn described his design for a compact marine engine, patented in 1841 (British Patent No.9072) as 'a novel disposition and construction of some of the working parts of steam engines, whereby such engines may be rendered more compact and more applicable to marine and other purposes'.

[above] In his patent *'Certain Improvements in Machinery used for Propelling Vessels by Steam'* (British Patent No.10,095 1844) he proposed novel gearing for the interface between engine and propeller shaft – the teeth on the driving wheel being on the inside rather than the outside of the wheel, increasing contact between wheel and shaft which he believed reduced stress on the teeth.

THE RAILWAY STEPHENSONS

[opposite page] The 1925-built replica of the 1837 2-2-2 broad gauge locomotive *North Star* which was built for the GWR by Robert Stephenson & Company. It was the GWR's first locomotive and thus had great historical significance. The replica is displayed at *STEAM, the Museum of the Great Western Railway* in Swindon.

The writer of *Notable Sights and Scenes England and Wales* published in 1900 by Cassell & Co. wrote 'The little village of Wylam, almost on the bank of the Tyne, is one of the oldest and most miserable-looking of all the colliery villages in Northumberland; but also one of the

GREAT WESTERN LOCOMOTIVE, 1842

MESSRS. STEPHENSON AND CO., NEWCASTLE-ON-TYNE, ENGINEERS

(For description see page 457)

[left] An illustration of *North Star* from the journal, *The Engineer*, dated 27 May 1892. The accompanying text suggests the locomotive was built in 1842, and offers a profile quite different to the replica at Swindon. In fact, this is the locomotive's profile after it was rebuilt with its wheelbase extended by 12ins in 1854.

[below left] George Stephenson's birthplace, photographed by Aberdeen-based George Washington Wilson. Wilson's photograph has been over-painted with figures and a speeding North Eastern Railway locomotive, and it was illustrated in the book *Notable Sights and Scenes England and Wales* published in 1900 by Cassell & Co.

connections with:
- James Watt
- William Fairbairn
- Henry Maudslay
- Isambard Kingdom Brunel

[right] Liverpool's Crown
Street Station, from a
coloured lithograph by
Henry Pyall (1795–1833).
It was closed in 1836 when
Lime Street Station opened.

[below left] One of the
National Railway Museum's
two replicas of *Rocket*, seen
here in 1980 preparing for
its run in the recreation of
the Rainhill Trials which had
taken place on the Liverpool
to Manchester Railway
between 6 and 14 October
1829. London's Science
Museum has the original
locomotive on display, albeit
as later modified.

[below right] The Olive
Mount Cutting on the
Liverpool and Manchester
Railway, seen here in 1904,
originally had two tracks and
opened in 1830. Designed
by George Stephenson, it
was widened in 1871 to take
four tracks into Liverpool
Lime Street Station.

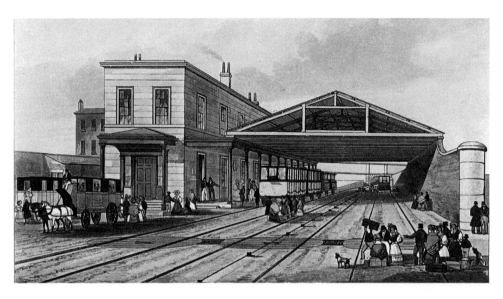

most famous, as the birthplace of George Stephenson, the poor collier lad who has re-made the
world, and who, when he died at the age of 56, was honoured by a public funeral and a grave in
Westminster Abbey'. His legacy, and that of his son Robert, can be seen the world over.

There is a fascinating building standing just a few yards from the present-day Didcot
railway station which recalls the formative years of Britain's railway network. Isambard
Kingdom Brunel's broad-gauge Great Western Railway reached Didcot in 1839, but a station
was not opened there until five years later.

Choosing to build his railway to a seven-foot wide gauge was a controversial choice by
Brunel, but from an engineering point of view, it was simple logic to him – wider rails allowed
wider carriages, more powerful engines and a much more stable ride. He obviously expected
other railway companies to follow his lead, but most opted for the 4ft 8$^1/_2$ins (1435mm) gauge
favoured by his engineering contemporary – and later friend – George Stephenson (1781–
1848), leaving Brunel's Great Western Railway rather isolated and in something of a quandary
if it was to play its part in the expanding the business of rail freight.

Recognising that there had to be a facility for transferring freight from one gauge to
another, the building known as the 'Transfer Shed' – which English Heritage dates to c.1840

with 20th century modifications – had tracks of both gauges. Just how great the difference between the GWR's original seven-foot gauge and what Brunel mischievously referred to as Stephenson's 'narrow gauge' can be fully appreciated when seeing the two sets of track side by side. Inside the transfer shed, dual-gauge track – the two gauges share three rails – ensured that whichever gauge the train ran on, it could unload its freight adjacent to the same platform face.

More than half of the world's railways are built to what used to be widely referred to as 'Stephenson Gauge', the standard gauge devised and built by George Stephenson, the father of the railway age. Together with his son Robert, (1803–1859) he initiated a railway-building era which, amongst other things, vastly increased the speed at which goods could be moved between major centres which had previously been determined by coaches and horses on road, or horse-drawn barges on the canals.

Just as the canals had done in the century before, railways revolutionised the movement of goods across the country. But while canals have remained virtually unchanged since they were constructed, railways have evolved and expanded to meet growing demand ever since steam traction was introduced almost two centuries ago. Given the progressive redevelopment

[above] George Stephenson, from a 1904 LNWR postcard.

[above left] George Stephenson's 54-feet long 'Skew Bridge' over the approach tracks to Rainhill Station – one of 16 skew bridges on the Liverpool and Manchester Railway. It carries the main road over the railway at an angle of 34 degrees and was the first in the world to go over a railway at such an angle. Work on it started in 1828 and took over a year. It was widened by four feet in 1963 to ease traffic flow.

[left] HST125 No.253-034 leaving Rainhill Station under the Skew Bridge during the re-enactment of the Rainhill Trials in 1980. The highly successful HSTs were introduced in 1976, originally planned as a 'stop-gap' until the intended introduction of the tilting Advanced Passenger Train. The APT project was abandoned in 1986. The last of the HSTs is scheduled to be withdrawn from service in 2024.

[above] A train approaching Manchester's Liverpool Road Station across the Water Street Bridge. Henry Pyall's 1833 lithograph is based on an original watercolour by Thomas Talbot Bury (1811–1877). An earlier lithograph was published in 1831 by Rudolf Ackermann. This 1833 version shows several differences from the 1831 original impression of Bury's watercolour – most significantly, gas street lighting has been introduced.

[above right] The engine house and 'Moorish Arch' at the top of the incline out of Lime Street Station. The cables can be seen between the rails. The arch was demolished when the cutting was widened in 1864, which coincided with the advent of more powerful locomotives.

which has taken place, it is remarkable that what is now celebrated as the oldest purpose-built railway station in the world still survives.

Liverpool Road Station in Manchester was opened in 1830 during the reign of King William IV by the Liverpool & Manchester Railway, and designed at street level to look like a terrace of elegant late-Georgian houses.

The route from Liverpool to Manchester had taken several years to come into being, the first route proposed and surveyed by George Stephenson in 1825 having failed to go ahead due to objections from the landed gentry along the way – the Earl of Derby and Lord Sefton both had interests in the canal network which they feared would become less profitable if the railway was built.

The second route surveyed by the eminent engineer Charles Blacker Vignoles in 1826 met similar objections. With George Stephenson soon back in the picture, however, a third route was finally agreed, and an enabling Act of Parliament passed in 1829.

Manchester's Liverpool Road Station is often described both as 'the world's first railway terminus' and 'the world's oldest railway station' – the second claim is true, the first is not – and it occupied a site on the corner of Liverpool Road and Water Street, incorporating a Georgian house built around 1810 on the corner itself, which became the Station Manager's, or Agent's, house and office.

The first passenger terminus on a locomotive-hauled railway was actually in Liverpool – Crown Street Station, designed by George Stephenson, opened in 1830 and closed just six years later. It was too far from the city centre whereas its replacement, Lime Street – still the terminus today – was at the heart of the 19th century city.

As might be expected for something so new, the construction of the Manchester terminus was covered in the local press, but at that time the term 'railway station' was not yet in common usage. So, in its 22 June 1830 edition, the *Manchester Mercury* used a term to which its believed its readers might more readily relate – describing it as a 'coach office'.

> 'Workmen are now employed in digging the foundations for a handsome and extensive coach office, eighty feet in length, to be erected immediately adjoining the house lately occupied by Mr. Rothwell. The coach office will consist of three stories [sic!], two of which will be below the level of the railway and the third above it. The western front of the building will look into Liverpool Road along which it will extend; and the other front (consisting of one storey) will look upon the railway exactly opposite the warehouses now erecting.'

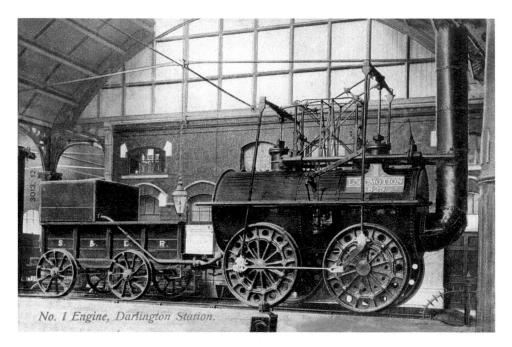

No. 1 Engine, Darlington Station.

[left] George and Robert Stephenson's *Locomotion No.1* – built in 1825 and originally named *Active* – was withdrawn from service in 1841 and subsequently used as a stationary engine. It was finally considered worthy of preservation in 1857. By 1892 it was on display in Darlington Station alongside Timothy Hackworth's 1845 0-6-0 *Derwent*. Both locomotives, now part of the National Railway Museum collection, are currently displayed at *Head of Steam*, the Darlington Railway Centre and Museum.

If the foundations were only just being laid in June, it is clear that the station would have been far from complete at the official opening of the railway on 15 September that same year. It being the early 1830s, of course, when the station did open there was a clear class divide inside the building – separate first and second class entrances at street level, separate booking offices and separate waiting rooms. There were even separate staircases up to the waiting trains – passengers being summoned by a bell when the time to board their train drew near.

There was also a small ticket office window at platform level which some have suggested was originally used for such things as issuing wages to the workforce, while the more commonly-held view is that it was for issuing tickets to third class passengers, offering them no protection from the Manchester weather whatsoever – but then again, as third class travel was in open carriages anyway, getting wet while buying a ticket would have been the least of those passengers' worries – but how they got up from street level to the trains is unclear. There was a staircase running up from the junction of Liverpool Street and Water Street, but little obvious sign of a route from there to the platform.

[below] The nameplate on the replica of *Locomotion*. It is displayed at the National Railway Museum's Shildon.

The station was sensitively restored in the early 1980s when it became Manchester's Museum of Science and Industry.

The warehouses are now home to everything from mill engines and railway locomotives to Manchester's earliest computers.

At the other end of the line an early challenge to the Stephensons and their contractors was the steep incline from Crown Street Station down to Liverpool docks and, later, to Lime Street Station. That was the lack of sufficiently powerful locomotives to climb back up the incline before heading east. The solution was innovative, if a little dangerous

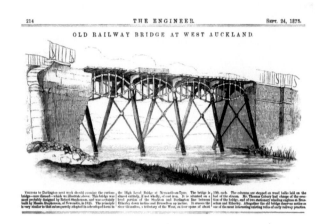

[above] George Stephenson's 1825 bridge over the River Gaunless on the Stockton and Darlington Railway was the world's first iron railway bridge, and the first 'lenticular truss' bridge, an idea later also used by Isambard Kingdom Brunel on his bridges at Chepstow, and Saltash.

[above right] The Camden engine houses and incline, one of 'eighteen plates' published in 1839 in Thomas Roscoe's book *The London and Birmingham Railway*, published by Charles Tilt.

[below] Robert Stephenson's 1828 *Lancashire Witch*, on a Tuvalu stamp from 1984. It was their first locomotive to have both a multi-tubed boiler and twin fireboxes.

– while trains ran down the incline controlled only by brakemen, they were rope-hauled back up by stationary steam engines housed in two imposing towers. Access between the two engine houses was by the stylish bridge which became known as the 'Moorish Arch'.

The first railway to use Stephenson's 'standard gauge' track was the Stockton and Darlington Railway, which was also the first in the world to use steam power, and the first to cross a river on an iron bridge – the lenticular truss bridge across the River Gaunless. The bridge was decommissioned in 1901 and preserved as part of the National Railway Museum's collection. It has been suggested that the innovative truss design which gave the bridge its strength may have informed Brunel's thinking when contemplating how best to bridge the River Tamar between Devon and Cornwall.

Once the decision had been made to use steam power rather than horses, a locomotive builder was needed. George and Robert rose to the challenge and formed Robert Stephenson & Company whose early locomotives would become legends – *Locomotion No. 1* on the Stockton and Darlington, and *Rocket* on the Liverpool and Manchester being the most famous. Both survive in preservation.

Interestingly, it was Robert Stephenson & Company to whom Brunel turned when ordering the first new-build locomotive for his Great Western Railway. The 1838-built 2-2-2 broad gauge locomotive *North Star* – their first for the GWR – obviously had great historical significance and was earmarked for preservation when it was retired in 1864. It was stored in a shed at Swindon intended to become the centrepiece of a future museum. But the best-laid plans sometimes don't work out, and it was cut up in 1906 to free up workshop space. A 1925-built replica is displayed at the STEAM museum in Swindon.

While the Liverpool and Manchester Railway became especially celebrated as a result of the publicity given to the 'Rainhill Trials' at which the Stephensons' *Rocket* demonstrated its reliability, their other projects arguably were more groundbreaking in the engineering challenges which had to be overcome.

Especially challenging was the London and Birmingham Railway, the contract for which George Stephenson & Son signed in 1830. To create the railway several long tunnels had to be excavated – including what was then the world's longest tunnel at Kilsby in Northamptonshire.

To cope with the steep mile-long incline from Chalk Farm, Camden, down towards Euston Road, a cable-hauled section and two large stationery engines in underground engine houses just north of the Regents Canal bridge had to be built – just as at Liverpool. Each engine drove a four thousand yard long endless rope,

[clockwise from far left] A ventilation shaft in Robert Stephenson's (1.35 miles, 2.2km) Kilsby Tunnel, Northamptonshire, on the London and Birmingham Railway (now the West Coast Main Line). The longest tunnel in the world at the time, it opened in 1838, three years before Brunel's Box Tunnel. Unstable geology caused construction problems and it took over three years to build.

On the same railway, Stephenson's Watford Tunnel was opened in 1837 and was just over a mile long, one of several tunnels the engineer was compelled to build as landowners did not wish to see the railway crossing their estates.

The castellated entrance to Stephenson's 777yd long (710m) Shugborough Tunnel on the Trent Valley Railway. It too was forced on Stephenson by a landowner reluctant to welcome the daily sight of trains passing by.

and all that was visible above ground was the pair of 130ft tall chimneys from the engine house and the ropes themselves running between the rails. Each of the ropes is said to have weighed in excess of 12 tons.

The journey out of the station was rope-hauled, the return journey relied on gravity and a very careful brakeman. Just why the cable-worked section was deemed necessary has long been the subject of debate. Some contemporary accounts suggest that steam working beyond Camden Town had not been authorised by the enabling Act of Parliament – yet there is clear evidence that locomotive-hauled trains did run right into the terminus. Other suggestions were that locomotives hauling heavy trains were insufficiently powerful to cope with the incline but, contemporary accounts dispel that.

Whatever the reason, the cable-hauled system operated from 1837 until 1844 when the used of banker engines replaced the cables for the uphill journey. Gravity continued to be used by trains coasting down into the station until 1857.

Interestingly, on conflict of interests grounds, George Stephenson was excluded from placing an order for locomotives for the London and Birmingham Railway with Robert's company, and the initial fleet of locomotives came from seven different builders, including Mather Dixon & Co of Liverpool, who had been established in 1825 as marine engine builders. They would go on to supply locomotives to several railways, including the Liverpool and Manchester.

George Stephenson was not just a railway man – he was a visionary and serial inventor – and long before he became the 'father of the railways', he was deeply concerned about the safety of miners in the sprawling coalfields of the north-east. He was the engine-wright at the Killingworth Colliery in Northumberland and developed his safety lamp along radically different lines to that which (unknown to him at the time) Humphrey Davy was developing elsewhere.

The Newcastle Literary & Philosophical Society was where he chose to demonstrate his prototype (*see page 12*) in 1815 – ironically in the same year in which Davy demonstrated his version in London. The two men had, independently, arrived at very different approaches to alerting miners to the problem of fire-damp in collieries – although Stephenson was apparently unwilling to consider the possibility of coincidence, and was particularly vocal in the accusations of plagiarism which he levelled against Davy.

Stephenson's lamp – known as the 'Geordie Lamp' was widely used in north-east collieries for most of the 19th century until the advent of electric lighting in mines, but the Davy lamp was the more widely used elsewhere, variations of its design continuing in use well into the 20th century.

The companies which the Stephensons established went from strength to strength. George Stephenson & Son excelled at railway construction, while Robert Stephenson & Company – founded in Newcastle in 1823 – enjoyed success for more than a century as major manufacturers of locomotives, having produced more than 3,000 before the end of the Victorian era. Their success continued into the 20th century with the building of new locomotive works in Darlington.

It is perhaps fitting that the most iconic of Robert Stephenson's bridges still survives – the cast-iron High Level Bridge across the Tyne which was completed in 1849 to carry the Newcastle and Berwick Railway railway north towards Scotland.

Amongst Robert's greatest achievements was the development, with William Fairbairn, of his 'tubular bridge'. As already discussed, two were built – at Conwy in (1849) and across the Menai Straits in (1850) – but after the fire which destroyed the Menai Straits bridge in 1970, the Conwy example is the lone survivor.

In developing the idea of the wrought iron tubular bridge, Stephenson called on Fairbairn's engineering genius in the development of the internal design of the box girder, the feature which gave the bridges their strength.

When the Britannia Bridge was opened, it was the longest wrought iron structure in the world, and its rigidity was down to a very clever piece of design. The tops and bottoms of

Photographed c.1902 with a locomotive traversing it, Robert Stephenson's 1,300ft long (408m) High Level Bridge over the Tyne at Newcastle was opened in September 1849.

[left] Stephenson's Britannia Bridge over the Menai Straits opened in 1850. This photograph, by Francis Bedford, dates from c.1857. The box construction of the railway bridge created immense strength – essential if it was to withstand the stresses of large heavy trains crossing it. The completed structure acting as a single fifteen hundred feet long girder supported by the three stone columns. The bridge was the first large construction project to benefit from hydraulic lifting gear, eight years before Brunel used them on the Royal Albert Bridge at Saltash in Devon.

[below left] An Edwardian postcard of the 'tubular bridge' across the river at Conwy, opened a year before the Menai bridge.

[below right] One of the hydraulic rams used in the construction of the Britannia tubular bridge.

each box section of every span were made up of a number of smaller box sections – eight at the top and six at the bottom riveted together – and their combined strength countered the tendency of the 'tubes' to sag under their own weight. The spans, each weighing more than 1,500 tons, were pre-fabricated on shore then floated out into the river and raised into position by massive hydraulic rams.

Afterwards, the rams were put into store in case they might be needed again. It turned out that they were not the only innovations from that project to be used elsewhere – and by none other than Brunel. The riveted box sections inspired the double-skinned hull design of the SS *Great Eastern*, and Stephenson's hydraulic rams were brought out of storage and, together with those supplied by the Tangye brothers, helped nudge the great ship towards the Thames on 31 January 1858. This was the last time the two men's paths crossed – Brunel died on 15 September 1859 aged just 53, and Robert Stephenson died just four weeks later on 12 October aged 55.

The Britannia Bridge was largely replaced after the 1970 fire, using the same stone piers – the box sections were completely destroyed – and in 1980 a roadway was constructed beneath the railway lines on the arched steel superstructure which was introduced during the rebuild. Such a structure would, of course, have been impossible back in the 1840s as the naval authorities had insisted that the railway tubes were high enough above the Menai Straits to allow tall ships to pass beneath.

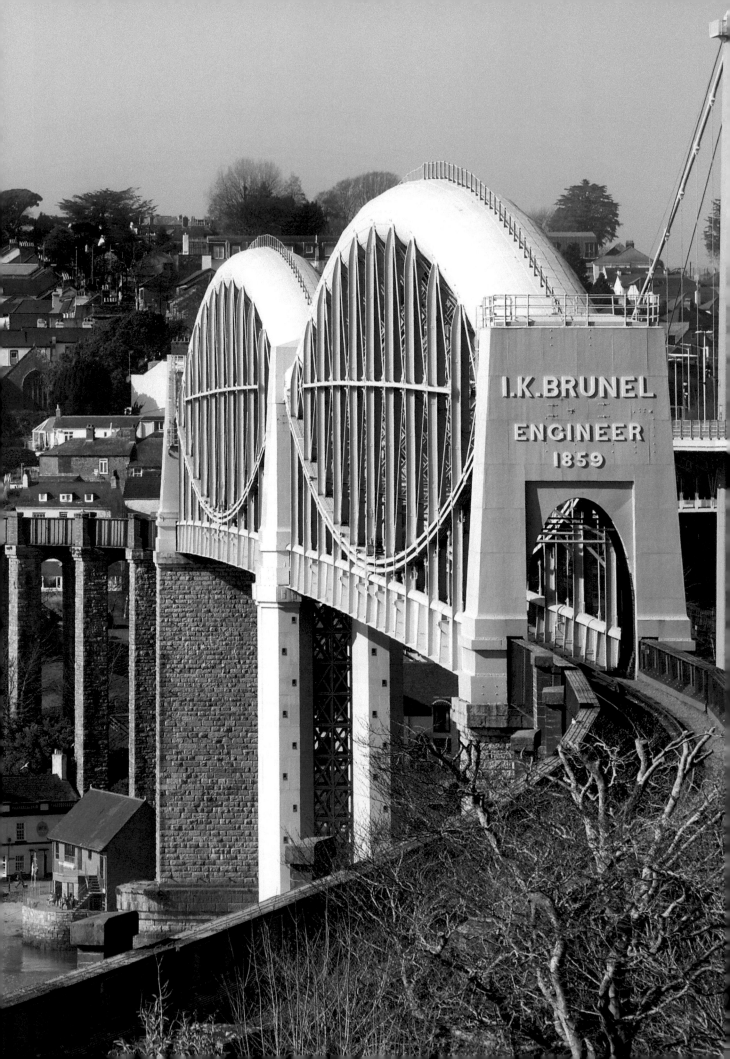

THE GENIUS OF MR. BRUNEL

Brunel's bridge across the River Tamar at Saltash is a masterpiece of engineering innovation – a beautiful structure which provided the first permanent railway link between Devon and Cornwall. Like many other projects in which he was involved, Brunel's bridge, developing an idea originally pioneered by George Stephenson, pushed the known boundaries of Victorian engineering to new limits. The bridge, opened by Prince Albert in 1859 and known as the Royal Albert Bridge ever since, is one of his finest achievements – a marriage of pioneering structural engineering and architectural elegance.

Since well before that event, however, Isambard Kingdom Brunel (1806–1859) had been leaving his unique mark on Britain. The bridge is just one of his many ground-breaking achievements, drawn from a career which revolutionised how people travelled. That so many of his achievements survive today – the Great Western Railway and the SS *Great Britain* to name but two – is a testament to just how 'right' his vision was.

The Saltash bridge is a 'lenticular truss' bridge – an idea introduced on 1825 by Stephenson – but turned 'upside down' by Brunel to give all the strength and stability of a conventional suspension bridge but without the need for tall towers. Its innovative design is in sharp contrast to the cost-saving design of the other bridges and viaducts which punctuated Brunel's broad-gauge line from Plymouth to Truro.

A proposal to build a railway from Plymouth to Falmouth – effectively extending Brunel's broad-gauge line to create a direct route from London to what was, at the time, a major port for transatlantic traffic – was first discussed in the mid-1830s, but it would be more than a decade before the first sod was cut. By the time that happened, Falmouth was already losing shipping trade to the fast-growing – and easier to reach – port of Southampton.

[above] Robert Howlett's image of Brunel standing in front of the launch chains of the SS *Great Eastern* was marketed as both a large format print and a small carte-de-visite for the Victorian family photograph album. Collecting celebrity photographs was a popular pastime even then.

[left & opposite page] The Royal Albert Bridge across the River Tamar linked Devon and Cornwall originally with Brunel's broad-gauge Great Western Railway and Cornish Railway. Alongside it is the modern suspension road bridge. The legend 'I. K. Brunel Engineer 1859' was added after his death that year, turning the bridge into a lasting and fitting memorial for the great engineer.

connections with:

- William Jessop
- William Fairbairn
- John Scott Russell
- The Railway
  Stephensons
- William Armstrong

Falmouth's growth had initially been aided by the simple fact that its deep-water port was closer to America, reducing the amount of coal which had to be carried by the steam packets which did the crossing. Carrying a few tons less coal meant that more cargo, or a larger number of paying passengers could be carried.

But as ships became larger and more efficient, that became less and less of a concern, with attention turning to passenger convenience, and the lower transportation costs of getting export goods to Southampton. The decision to go ahead, however, was influenced to some extent by the growing leisure market, as the projected growth in holiday traffic was thought likely to balance the loss of freight trade.

That would turn out to have been an optimistic assumption, resulting in the proposed Cornwall Railway being hugely under-capitalised.

When Steam Packet services from Falmouth were withdrawn in 1848, the railway project ground to a halt, and for four years no work was undertaken. Little had actually been achieved except for a detailed survey of the geology of the River Tamar's banks and riverbed.

Crossing such a wide expanse as the 1,100ft (340m) wide estuary – an important shipping lane at the time – coupled with the navy's insistence that any bridge must have sufficient clearance to let their tallest sailing vessels pass beneath it, presented a significant engineering challenge.

Brunel had originally intended to build a twin-track seven-span viaduct, with a wide central span and a clearance of 80ft (23m) at high tide. The Admiralty thought differently and required a clearance of 100ft (31m).

His second design had four piers but that, too, was rejected, with the naval authorities insisting that there should be just one pier in the centre of the river.

So the single-track line carried through the lenticular truss bridge which is still in use today was developed. The lenticular truss design is a very clever bit of engineering and is self-supporting – thus no horizontal stresses are transferred horizontally to the piers themselves. The strength in each of Brunel's spans came from a single tubular arc from which was 'suspended' the decking on which the railway track was laid. That design meant that the trusses could be constructed off site, and raised into position as completed structures. The construction yard was established – and extensively photographed – on the Devon shore.

The foundations of the first pier were laid in July 1853, but it was over four years later – on 1 September 1857 – that the first of the bridge's two double longbow-shaped trusses was floated into position between the Cornish coast and the central pier. Then, at a rate of six feet per week, it was slowly raised up into position using hydraulic jacks, with the shore pier being built up beneath it as the truss was raised.

That first truss was finally secured in position 100ft above the river ten months later on 1 July 1858. Less than two weeks after that, on 10 July, the north truss was floated out from the

bank and manoeuvred into position between the central pier and the Devon bank of the river.

Prince Albert agreed to it bearing his name, and he performed the opening ceremony on 2 May 1859, having travelled from Windsor on the Royal Train – Brunel's broad-gauge, of course.

While a great deal of money had been spent on the bridge, the many other viaducts needed to span Cornwall's many rivers and estuaries had to make do with cut-price structures as the Cornwall Railway was quickly running out of funds, and despite Brunel's warning that his low-cost design would have much higher maintenance costs than conventional viaducts, the directors commissioned a series of 42 hybrid bridges – stone pillars topped by wooden fantails on which the track was laid.

The woodwork was assembled at a specially built facility at Lostwithiel where the timber arrived by boat and was cut to size and treated.

The wood was preserved with the application of either mercuric chloride – known as Kyanising – or zinc chloride – Burnettising – named after their inventors in the 1830s, John Howard Kyan and William Burnett. The two longest of these hybrids were both built in Truro – the 20-pier Carvedras Viaduct and the 15-pier Moresk Viaduct.

In 1892, of course, the decision was taken to abandon Brunel's pioneering broad-gauge system, and replace it with standard gauge, but three years before that costly, but necessary, decision was taken, the Cornwall Railway's debts had started to overwhelm the company and in 1889 it had passed into the hands of the GWR which had partly funded it from its inception.

Long before then, however, Brunel's warnings about the cut-price viaducts had proved correct, the first of them having been replaced with conventional stone arch bridges as early as 1871. Most of the others were replaced in the early years of the 20th century. Despite continuing concerns about their durability, a few of these highly original viaducts survived into the 1930s – the last two to go being the 11-pier Carnon Viaduct in 1933 and the fourteen-pier Collegewood Viaduct in 1934.

In 1832, the 26 year-old Brunel was appointed to report on the state of William Jessop's Floating Harbour in Bristol where a relentless build-up of mud was giving cause for concern. The dock had only been operational for 13 years at the time.

Jessop had designed a dam with an overflow – or 'overfall' as he had called it – to keep the water level constant inside the harbour. Unknowingly, he had built in a problem more severe than anticipated. While recognising from the outset that the harbour would need to be regularly dredged, he had underestimated just how much mud would be drawn in by the tides, feeder streams and ship movements. As the mud built up, clear water was being sluiced over the top of the dam.

Brunel proposed a radical solution by cutting a deep channel through which faster flowing excess water could be sluiced well below the surface, taking much of the mud with it. Thus Jessop's 'overfall yard' became today's Underfall Yard. He retained his involvement with the harbour – as Consultant Engineer – for many years until other projects took over much of his time from the early 1850s

Sadly not in the public eye, and also in Bristol, is one of the earliest-surviving examples of his innovative bridge designs. No longer in use, and tucked away alongside the huge 1965-built Plimsoll Bridge – also a swivel bridge – sits Brunel's first wrought-iron swivel bridge which was built in 1849 to offer vehicle access across the south entrance lock of the floating harbour. The bridge was constructed at the same Great Western Steamship works in which the SS *Great Britain* had been built a few years earlier, and where the restored ship now sits today.

[above left] The 15 piers of Brunel's Carvedras Viaduct in Truro still stand alongside its 1902 arched stone replacement. By the time this view of load-testing the bridge was published as a postcard, it had already been dismantled and replaced.

[above right] The 14-arch 954ft (291m) long and 100ft (30m) high Collegewood Viaduct just south of Penryn Station was the longest of Brunel's wooden fan-arched viaducts in Cornwall. It survived until 1934, becoming the last of its type to be replaced by a conventional stone-arched structure.

[opposite top] A driver's view of the single track line across the Royal Albert Bridge, Saltash – from a photograph privately published as a postcard c.1905.

[middle] One of the hydraulic rams used to raise the tubes of the Royal Albert Bridge into place.

[bottom] The Saltash Bridge construction site on the Devon shore, photographed between 30 August and 6 September 1858. The photographer's name is not recorded.

The wider entrance lock which it bridged was also designed by Brunel, built in 1843–44 to replace William Jessop's original and thus enable the SS *Great Britain* to pass out of the floating harbour after completion.

Originally 120ft long (36.4m) and weighing over 70 tonnes, the wrought-iron bridge, which had a wooden deck, was initially hand-cranked like early railway turntables and spanned Brunel's lock until 1872 when it was shortened by 10ft (3m) and relocated over Howard's Lock.

Hydraulic power was introduced into the docks powered by two 44hp steam engines in a pumping station designed and built by William Howard. The hydraulics were by Sir William Armstrong, and when electricity replaced steam, Armstrongs modified the system. A new pumping station at Underfall Yard with motors by Fullerton, Hodgart & Barclay of Paisley opened in 1907.

Brunel's bridge was finally taken out of service when the Plimsoll Swing Bridge opened in 1965, and when that bridge is swung open to allow vessels to pass through, Brunel's bridge sits just four inches below it.

The massive lock gates themselves were designed with sealed air chambers in their lower sections, to give them a degree of buoyancy, thus reducing the friction on the wheeled mechanism on which they were carried – and reducing the energy used to open and close them.

Between the completion of the Bristol bridge and work starting at Saltash, Brunel designed what became known as the 'Great Tubular Bridge' to carry his Great Western Railway across the River Wye at Chepstow.

Like the later Saltash Bridge, the Wye bridge was a suspension bridge, and for many decades these were the only two suspension bridges on the railway network.

The railway deck was suspended from a tensioned slightly bowed tubular truss and is it widely seen as the 'test-bed' for the much wider spans at Saltash. It was built on a site adjacent to the river by Finch and Willey of Liverpool, who later became major bridge- and ship-builders in the area.

The six feet diameter tubular truss was constructed in cast iron and wrought iron, its key design feature being the bowed and tensioned tube, made of overlapping hand-shaped iron plates one and a half inches thick and riveted together. Shaping metal plates of such a thickness by hand in the days before powerful steam hammers must have required considerable effort from Finch and Willey's workforce. The tube's strength came from the large circular bulkheads set at intervals within its 309ft length – the same technique which had been used on the Bristol Swivel Bridge.

Brunel never 'ran with the crowd', and his railways were a notable example. 'Standard' gauge was already becoming well established when Brunel chose not only to go with a broader gauge, but to go with a broader gauge than any of the other broad gauge lines already under development.

[left] The trackbed of Brunel's prototype tubular truss bridge – the railway bridge over the River Wye at Chepstow – was slung from the self-supporting overhead tubular trusses. It was partially rebuilt 11 years later.

[below left] A short riveted section of the 1852 tubular truss from the Chepstow Bridge has been preserved and now stands in a modern housing development up the hill from the present bridge.

[below right] Seen here at low tide, the 1962 railway bridge which replaced the tubular bridge is still carried on Brunel's 170-year old riveted cast iron columns. The 1962 underhung replacement truss is visible beyond the columns.

His rationale was entirely logical – the broader gauge lowered the centre of gravity of the vehicles – both locomotives and rolling stock – making for a much more spacious and stable vehicle.

The wider carriages allowed less cramped seating, and that lower centre of gravity significantly reduced the tendency of a train to tilt as it went round corners. He therefore correctly asserted that his railway could deal more safely with curves when trains were travelling at speed. As his locomotives could accommodate larger boilers thanks to the wider frames on which they were mounted, they were much more powerful than those being used

[opposite page, from the top] The internal structure of Brunel's 1852 Swivel Bridge, which originally had wooden decking. It was formerly sited over his South Lock entrance to Bristol's Floating Harbour.

A profile view of the bridge 'beached' sitting partly below the modern Plimsoll Swing Bridge. A close copy of the bridge, built in 1856, spans the entrance to the Petrovsky Dock at the former Russian naval base in Kronstadt near St. Petersburg and, restored recently, is still operational.

Currently under restoration, this image shows the tubular construction of the bridge, and the wheel system which ran in a circular track as the bridge was being rotated.

Brunel's South Entrance Lock with the Plimsoll Bridge passing over it.

[right] A Great Western express leaving the east end of Brunel's 1.8 mile long Box Tunnel, c.1910. The arch to the right of the tunnel gave access to the quarries – opened in 1844 – from which much of Bath stone was excavated. The quarry tunnel was later used as a munitions depot during the Second World War, a communications centre during the Cold War, and is now a secure document storage facility. When opened, it was the longest railway tunnel in the world, almost two-thirds of a mile longer than Robert Stephenson's Kilsby Tunnel which had opened three years earlier.

[below] The early morning sun shining through Box Tunnel on the Great Western Railway's main line on 9 April 2017 – taken when the line was temporarily closed and no trains were travelling through while electrification work was underway. The wide space between the two sets of track is a reminder that the tunnel originally accommodated Brunel's broad gauge railway. (*image courtesy GWR*)

on other railways, so higher speeds were part of the plan and became a heavily promoted feature of the railway.

But that came at a cost – at seven feet between the rails, the track was more difficult to lay and had a tendency to splay unless fixed on to a solid wooden foundation – known as a 'baulk' – running continuously under each rail. He later improved stability on curves yet further by widening the gauge by a quarter of an inch.

The Great Western Railway was a remarkable achievement when it opened, and is still considered a masterpiece today. But Brunel was a pragmatist, realising early on that Stephenson's 'standard gauge' was being much more widely adopted and that an easy way to

transfer goods from one gauge to another was essential if the Great Western was to become a dominant player in the evolving railway freight network.

An interchange between the two gauges was established at Didcot in Oxfordshire, where goods could be transferred from the London to Bristol broad gauge line to standard gauge routes north. The unique wooden Transfer Shed which was built there – dated by English Heritage to c.1840 with some 20th century modifications – enabled the easy transfer of freight, under cover, while the complex track layout enabled both standard and broad gauge trains to access the shed's platform.

The shed formed part of the original Didcot Station buildings, and was only relocated to its current position as a key exhibit in the Didcot Railway Heritage Centre in the early 1980s.

The Great Western set new standards for construction along its entire length, but Box Tunnel must be considered one of Brunel's greatest achievements. Its design called for a tunnel almost two miles long, perfectly straight, cut through some challenging geology. Few thought it could be built, and more than 100 men lost their lives during the three years of its construction.

Some described the idea as 'dangerous, extraordinary, monstrous and impractical'. Nobody had ever built a tunnel like it before, and yet the great man's achievement is still an essential feature of the GWR line today.

It was said to be so perfectly straight that on a certain date, the rising sun would shine right through it. And since an article appeared in the *Devizes Gazette* the year after the

[below] The drive wheels of the replica *Iron Duke*, the original having been built in 1847 at Swindon.

[bottom] Just inside the 'STEAM' museum in the former GWR works in Swindon, 'Brunel' stands with the surviving driving wheels from Daniel Gooch's Swindon-built *Lord of the Isles*.

tunnel opened, that date has been believed to have been 9 April, Brunel's birthday. Just coincidence, or one of Brunel's deliberate intentions? Recent research, however, has suggested that the phenomenon probably lasts for several days, first occurring a few days earlier, on 6 April, which happened to be the birthday of Brunel's oldest sister Emma Joan. Either way, the tunnel is precision engineering on a massive scale. With the interior of the tunnel blackened by years of smoke and soot, however, it was difficult to prove or disprove the story – until, that is, the tunnel was closed for several months in 2017 while the line was being electrified.

By the time Box Tunnel opened in 1841, Brunel was already on to his next project – creating the biggest and most luxurious steamship yet built – SS *Great Britain*. He had designed the vessel as a paddle-steamer, as had been his SS *Great Western*, but during construction he became fascinated – then impressed and finally convinced – by the potential of the screw propeller.

Not dissuaded by the fact that the boat was already under construction, Brunel persuaded his directors to back his decision to turn the engine through 90°, add a chain drive from the large wheel on the crankshaft down to a smaller rachetted wheel on the propeller shaft, to which was fitted a large iron propeller. The specially designed system allowed the propeller to be raised out of the water when the ship was travelling under sail, and lowered back when under steam.

The engine, however, was very inefficient, operating at a pressure of just five pounds per square inch and using a tubeless

boiler which was really little more than a giant kettle. Notwithstanding its inefficiency, it fascinated travelling passengers who looked forward to the Chief Engineer's mid-Atlantic tours of the engine room. It was replaced in 1852 by a smaller but more powerful engine. Tours of that proved just as popular.

Brunel's decision to build the ship out of iron was, typically, breaking new ground. The first passenger vessel with a riveted iron hull – the horse-drawn canal boat *Viking* – had been launched just 22 years earlier in 1819 at Calderbank on the Monklands Canal in Coatbridge near Glasgow, and Brunel was planning to build what would be, at the time of her launch in 1843, the biggest ship in the world by some measure.

The story of the SS *Great Britain* is a story of success, failure, abandonment, rescue and restoration. Now preserved in her home port of Bristol, the vessel offers a unique window into the early days of transatlantic steamship travel. But it might have all ended differently.

On 26 May 1886, having already lost sections of both her fore mast and main mast in a ferocious storm – and no longer a steamship having had her engine removed and been returned to sail – the *Great Britain* ran aground off Port Stanley harbour in the Falkland Islands, her sailing days over, just 40 years since her maiden voyage had started from Bristol in spring 1845.

She had been a revelation to all who saw her set off – the world's first transatlantic liner to be fitted with a propeller rather than paddles, and the biggest steamer the world had ever seen. She offered unprecedented luxury and, thanks to her massive 1,000hp 4-cylinder engine, unprecedented speed. But while luxury may have been enjoyed by the first class passengers, other classes of passengers, together with the crew, were consigned to accommodation deep in the bowels of the largest steamship in the world where life was very different.

Working temperatures in the cramped boiler room often exceeded 50°C, and the stokers each shovelled around a ton of coal per shift. Life for the engineers would not have been much more comfortable – their cabins were sited just a few feet away from the constant noise and vibration of the engines. Despite the difficulty of sleeping with all that noise, they were required to be alert throughout their long shifts on duty during the 14-day crossing of the Atlantic.

After running aground in 1886, the ship was used as a floating store for more than 50 years before finally being abandoned in Sparrow Cove in 1937 – by which time her revolutionary iron hull had been in the water for almost 94 years. Thirty years passed before the idea of repatriating her and restoring her was first discussed, with the hulk eventually arriving back in Bristol in June 1970 and being floated back into the Great Western Dock where she had been built.

Forty years of stabilisation and restoration followed. While some of the interior has been recreated, enough has been left almost 'as found' to allow visitors to explore the structure of the world's first 'modern' passenger liner.

The SS *Great Britain* may be the only surviving example of one of Brunel's steamships, but tantalising glimpses of the scale of his last and largest vessel – the largest ship ever built in the Victorian era – can still be seen.

Close to the water's edge on the north shore of the Thames at Millwall, the array of wooden posts and beams, uncovered a few years ago during redevelopment are all that remains of the slipway from which Brunel's SS *Great Eastern* was launched.

Here was yet another example of Brunel the visionary. More than 50 years before the ill-fated RMS *Titanic* sank in 1913, Brunel's *Great Eastern* – the vessel was originally to be called *Leviathan*, a name befitting the biggest ship the world had ever seen – had been conceived and built with a double-skinned hull, the very feature which might have saved *Titanic*.

Everything about *Leviathan* was on a colossal scale. She was to be almost 700ft long and 18,915grt – massively bigger than the 322ft long *Great Britain* at 3,650grt.

The steam engines which drove the 56ft diameter paddle wheels were built at John Scott Russell's shipyard adjacent to the construction site, while the propeller was driven by an engine built by James Watt & Company at their Soho Works in Smethwick. That company had changed its name from Boulton & Watt in 1849 in honour of its founder, the pioneer of steam.

Russell – an eminent engineer in his own right and Brunel's original partner in the *Leviathan* project – had proposed building a massive new drydock in which to build the ship, but Brunel thought the cost of so doing was excessive, resulting in the decision – later to prove problematic – to assemble the vessel side-on to the Thames at Napier's Millwall yard next door to Russell's.

Russell and Brunel had successfully worked together on two of Brunel's other ships – the SS *Victoria* and the SS *Adelaide*, both completed in 1852 and of similar size to the SS *Great Britain*, but the scale of the *Leviathan* project would bring their partnership to an abrupt end.

On 3 November 1857 a crowd of many thousands had turned up, and a photographer was ready to take an historic picture at 12:30pm as the biggest ship in the world slid sideways into the river.

Henrietta Hope, the daughter of one of the major investors in the vessel, christened her *Leviathan*, but the ship moved just a few feet and stuck fast. That was much to the surprise of everyone – including Brunel – as it had been widely expected that she would be named *Great Eastern*, and equally widely expected that she would have slipped effortlessly into the rising tide.

Her name was changed to *Great Eastern* in July 1858 six months after she was eventually floated out on January 31 1858, aided by Tangye Brothers' hydraulics.

[opposite top] The 2005 replica of Daniel Gooch's 2-2-2 *Fire Fly* standing in the Transfer Shed at Didcot. The original locomotive was built at Jones, Turner & Evans's Viaduct Foundry in Newton-le-Willows, Lancashire, in 1840, the same year the Transfer Shed is believed to have been built.

[opposite bottom] The recreation of the boiler room on Brunel's SS *Great Britain*.

[below] Brunel's SS *Great Britain*, in the Great Western Dry Dock in Bristol's Floating Harbour from where she had been launched in 1843.

[right] The chain-drive engine was originally designed to turn paddle wheels but Brunel elected to turn it through 90 degrees and use it to turn a propeller shaft instead.

[below] The cavernous interior of the hull, photographed in 1983 with the early stages of restoration under way.

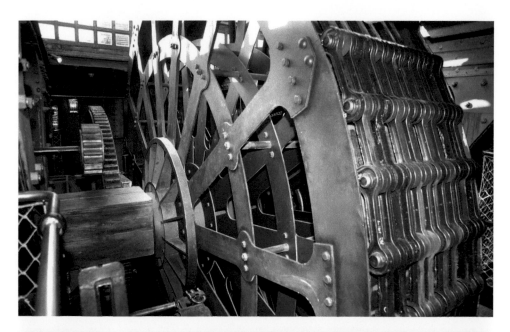

Birmingham-based Tangye brothers, were asked to design and build powerful hydraulic rams to help launch the ship. and Richard Tangye later commented that 'we launched the *Great Eastern*, and she launched us', but that was not the full story. In his autobiography, *One and All*, Tangye explained what actually happened:

'One dark evening in the winter of 1856, Brunel's agent came to our little workshop, which was down an entry behind a baker's shop, and rang our bell. I

[clockwise from top left] A first class cabin, as recreated in the restored ship.

The recreation of the passengers' dining saloon is not just for show – as a key part of its business plan, the restored ship has become a popular venue for events, conferences and weddings.

SS *Great Britain*'s massive propeller and rudder

The stern decoration is reminiscent of sailing ships from the pre-steam era.

opened the door, but the gentleman apologized, saying he had made a mistake, and was moving away; but I could not afford to lose a possible chance of business and so said "Whose place are you looking for, sir?" He replied "Tangye's." I told him he had come to the right place, and invited him in, when, having told us his business, he was quickly reassured upon seeing the machines before him. He ordered several of them, but my brother told him they would not be sufficient to effect the launch, with which he did not agree.'

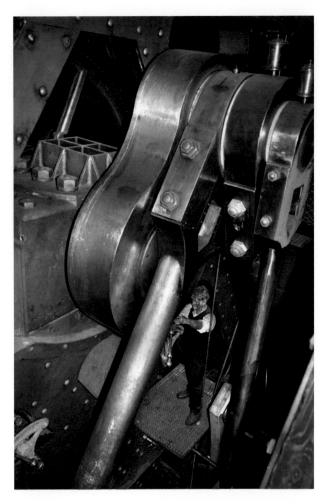

The Tangyes were proved right when the ship failed to launch and were concerned about the impact that failure might have on their business. Several decisions had been forced upon Brunel due to cost constraints, and it is possible that ordering too few rams was driven by that same financial pressure. One significant decision had been refusing to consider the cost of building a new dock in which to build the vessel, another had been several reductions in the total construction budget – which had started out at £500,000 but was incrementally reduced to under £300,000.

Then Brunel's plan to build a single 700ft wide slipway had to be abandoned – again on cost grounds – and replaced with the much cheaper option of two slips, one under the bows, the other towards the stern.

The engine in Brunel's SS *Great Britain* operated at a pressure of less than five pounds per square inch, using a tubeless boiler which was little more than a giant kettle. It fascinated passengers who looked forward to the Chief Engineer's mid-Atlantic tours of the engine room. The replica engine in the restored ship is turned by electricity for the benefit of visitors.

[above right] SS *Great Britain* tied up at Brandon Wharf in Bristol's Floating Harbour in 1844 while being fitted out. Long attributed to William Henry Fox Talbot, this image is now thought to have been taken by amateur photographer, the Reverend Calvert Richard Jones.

After the failure, Brunel was advised by Robert Stephenson, to get Tangyes to double the number of hydraulic rams, while Stephenson had brought out of storage the massive steam ram which he had used in the late 1840s to raise the tubes on the Britannia Bridge across the Menai Strait. Together, they were successful. Brunel is said to have paid the Tangyes more than £70,000 for their rams – equivalent to about £28M today.

Richard Tangye, the businessman in the family – the other brothers were the engineers and innovators – had done a deal with Brunel whereby the additional rams had been provided at a special price if the earlier debacle, and the importance of the contribution made by Stephenson's ram, was played down.

The launch problems had arisen because, as built, the twin wooden slipways were not quite at the same angle to the river, causing the ship to skew slightly as it started to move.

For everything to go as planned, both slipways should have been exactly the same height and with exactly the same gradient, as indeed they had been when first built. But the concrete foundations which supported the structures proved not to be robust enough as the weight of the ship increased during construction and, due to progressive settlement of the river bank over the almost four years of construction, one of the slipways had become steeper than the other.

During redevelopment of the Millwall waterfront a few years ago, the launch ramps at the Napier Yard were rediscovered, and at low tide the concrete and stone foundations can still be seen running into the river. Examination of the site has revealed that, while Brunel planned

In 2010 the dock was sealed around the ship with a 'glass sea' and visitor access to the floor of the dock was enabled, allowing close inspection of the elegant clipper-shape of the riveted iron hull. By that time, 35 years exposed to Bristol weather had done more damage to the hull than all her years in the water so, to preserve the hull plating – which is now paper thin in places – from further deterioration, warm air is circulated around it and sensors constantly monitor its condition.

for a 1-in-12 gradient, the bow ramp actually had a gradient of 1-in-11, with the stern a gentler 1-in-15 – a significant discrepancy.

The *Great Eastern* turned out to be Brunel's last great project. After years of deteriorating health, he suffered a major stroke on 5 September 1859 and so he never saw the *Great Eastern* set off on towards Weymouth the following day, in preparation for her planned maiden voyage across the Atlantic. He died eight days later. News did reach him, however, of the great boiler explosion on 9 September which had scalded five stokers to death, and rectifying the damage postponed the ship's first transatlantic voyage into the early summer of 1860.

[left] From a contemporary print, the launch of the SS *Great Britain* on 19 July 1843. She set sail on her maiden voyage two years later in July 1845. At that time, and for the next nine years, she was the longest passenger ship in the world.

[below] The replica maker's plate on the restored vessel.

[right] *Frank Leslie's Illustrated Newspaper*, an American journal, published a special supplement on 17 September 1859 to mark the *Great Eastern*'s first transatlantic crossing. The supplement illustrated the construction of the great ship. This view shows the pioneering double-skinned hull during construction.

[below] One of the two launch ramps for the SS *Great Eastern*, revealed during redevelopment of the waterfront. As built, the wooden trestles would have been topped with iron rails on which the ship was intended to slip into the Thames.

Part of Brunel's original vision had been of a ship big enough to make round trips to Australia without re-coaling, but the discovery of plentiful supplies of coal in Australia made that capacity unnecessary.

Never achieving her planned success as a passenger liner, it was in the much less glamorous but hugely important task of laying transatlantic cables that the ship achieved commercial success. It was her sheer size which made her an obvious choice to undergo conversion for use as a cable-laying ship. With funnel No.4 and two of her boilers removed, her passenger accommodation stripped out and replaced by three huge 'cable tanks', the heavily modified ship eventually successfully laid the second transatlantic telegraph cable in 1865–6, and several other cables thereafter. No other ship in the world had the capacity to carry the 2,600-mile long cable. She had finally found a role which she could undertake profitably, and was used to lay cables across the Indian Ocean as well as the Atlantic.

In a curious way, Brunel's last structure to be built was also his first, and it was certainly audacious – laying down a marker for the innovation which would later become the hallmark of his life's works. In his biography of his father, Isambard Brunel noted

> 'during the greater part of 1829, Mr. Brunel kept himself fully employed in scientific researches, and in intercourse with Mr. Babbage, Mr. Faraday, and other friends, but he was without any regular occupation'

Until, that was, his first bridge commission had come along in 1831 – spanning the 700ft wide Clifton Gorge at Bristol with a suspension bridge. It was the first of his many daring propositions, but one which he would never see through to completion. Bridging the gorge was immensely challenging and the project was seriously under-funded from the outset. To try and ameliorate the impact of that funding shortfall, the committee overseeing the commission made repeated demands for cheaper ways to be found of completing it. After seven years – the foundation stone had been laid on 27 August 1836 – work ground to a halt in 1843 and the bridge would not be completed until 1864, several years after Brunel's death.

One of the ways in which they tried to persuade him to cut costs was by replacing the chains slung between the towers with cables, which would have considerably reduced the

[above] Brunel's first transatlantic steamer, the SS *Great Western* was built in Bristol in 1837 and powered by a two-cylinder low pressure steam engine built by Henry Maudslay. At the time of her first Atlantic crossing – from Bristol to New York in April 1838 – the 1,340grt steamer was the largest passenger ship afloat. She ended her days in 1856 as a Crimean War troop transport.

[left] The SS *Great Eastern* under way – from a contemporary chromo-lithograph.

THIS BRIDGE
WAS DESIGNED IN 1830 BY
ISAMBARD KINGDOM BRUNEL
(1806-1859)
CONSTRUCTION BEGAN IN 1836 BUT WAS
INTERRUPTED IN 1843 THROUGH LACK OF
FUNDS. IT WAS NOT UNTIL 1864 FIVE YEARS
AFTER BRUNEL'S DEATH THAT THE BRIDGE
WAS COMPLETED AS A MONUMENT TO
HIS FAME. THE CHAINS USED BEING THOSE
FROM THE HUNGERFORD BRIDGE DESIGNED
AND ERECTED BY HIM IN 1843.

weight of the structure. He refused, believing that to be an unacceptable compromise which would weaken the bridge and shorten its working life. His wish prevailed and the bridge is still carrying road traffic today, nearly 160 years after it was eventually completed. He had argued that in Britain's damp climate, cables would last no more than 50 years, a fact which has subsequently been borne out by progressive cable failure on a number of modern suspension bridges.

It is remarkable to think that at the same time as he was working on the Clifton Bridge, Brunel was also engaged in several other major projects – most significantly the Great Western Railway, authorised by Act of Parliament in 1835 when he was just 29 years old, and, Just as surprising is the fact that the GWR was running its first trains just three years later.

A quotation from 'an old friend' in the obituary published by the Institution of Civil Engineers summed him up admirably, albeit in a manner with which those who had crossed swords with him would take issue – for he cannot have been an easy person to deal with, expecially for those who might, at some point, have proposed alternative ways of doing things.

'In youth a more joyous kind-hearted companion never existed. As a man, always over-worked, he was ever ready by advice, and not unfrequently, to a large extent by his purse, to aid either professional, or private friends. His habitual caution and reserve made many think him cold and unworldly, but by those who saw his exterior only, could such and opinion be entertained. His carelessness of contemporary public opinion, and his self-reliance on his own character and that of his works, were carried to a fault He was never known to court applause. Bold and vigorous professionally, he was as modest and retiring in private life. He was cut off in his fifty-fourth year, just when he had acquired the judgment which, in such a profession as that of the Civil Engineer, can only be obtained by long practice and experience, and when the greatest work of his life had reached the very eve of completion'.

[above left] One of Francis Bedford's photographs of the Clifton Suspension Bridge under construction. Brunel had died long before work on the bridge reached this stage. Work to complete the project was undertaken by his friend and one-time collaborator on the Great Western Railway, the engineer Sir John Hawkshaw (1811–1891) who, with William Barlow, completed the bridge in 1864. John Hawkshaw's name ought to be much more widely known than it is. He built the Severn Tunnel – an engineering masterpiece – for Brunel, as well as building railways, canals, bridges, and harbours across the world. He advised on a possible route for the Suez Canal, and was an early promoter of the idea of boring a Channel Tunnel.

[above right and opposite] Brunel's bridge spans the Clifton Gorge nearly 300ft above the River Avon. It still serves the purpose for which he designed it in 1836.

[inset] A commemorative plaque on the tells of its troubled genesis.

James Nasmyth. (1808-1890)
Engineer, Industrialist, Entrepreneur, Artist, Astronomer.

MR. NASMYTH'S HAMMER

When Isambard Kingdom Brunel and Thomas Guppy decided to build the biggest steamship in the world, and build it out of iron rather than timber, the changes in scale and materials required them to meet and resolve a host of completely new engineering challenges.

Their first steamship – the SS *Great Western* – had helped demonstrate that passenger steamers could cross the Atlantic profitably, but Brunel reasoned that bigger ships would be even more profitable.

Being Brunel, of course, he didn't just plan a ship which would be a little bit bigger, he planned to follow the wooden-hulled and iron-braced SS *Great Western* – 250ft long and with a gross registered tonnage of 1,340 and a displacement of 2,300 tons – with the iron-hulled SS *Great Britain* which would be 322ft long and have a displacement of 3,400 tons.

When the SS *Great Western* was completed in 1838, steam power for ships was still remarkably primitive – massive simple single- and twin-cylinder low-pressure engines operating at no more than a few psi, turning large heavy paddle wheels. At the heart of the vessel's operation was a large forged iron paddle shaft.

[above] James Nasmyth, photographed in the 1860s.

[below] Before the steam hammer, foundries used water-powered tilt hammers, of which this rare survival at Finch Foundry in Sticklepath in Devon is a small example. Many small foundries continued to use their water-powered hammers rather than meet the cost of even a small steam hammer, but for large fabricators, the steam hammer greatly expanded their capabilities.

[opposite page] A restored Nasmyth hammer, built in 1851, stands beneath a protective canopy – bearing prints of some of Nasmyth's original drawings – at the entrance to the Nasmyth Business Park in Patricroft, the location of the engineer's Victorian foundry.

connections with:

• Henry Maudslay
• James Watt
• James Nasmyth
• William Fairbairn
• Isambard Kingdom Brunel

[left] James Nasmyth with one of his steam hammers, photographed in the early 1850s by his friend and fellow member of the Manchester Literary & Philosophical Society, Joseph Sidebotham.

[right] Nasmyth's hammer as it was exhibited at the Great Exhibition at the Crystal Palace in London's Hyde Park, in 1851.

In forges and foundries large water-powered hammers had long been used to beat the metal into shape. Scale the ship and its engines up so something of the size of the proposed SS *Great Britain*, and there simply were no available hammers large enough to help forge-masters create the drive shafts such vessels would need.

The steam hammer had first been suggested by James Watt in his 1784 Patent in which he proposed the development of:

> "Heavy Hammers or Stampers, for forging or stamping iron, copper, or other
> metals, or other matters without the intervention of rotative motions or wheels,
> by fixing the Hammer or Stamper to be so worked, either directly to the piston
> or piston rod of the engine".

This proposal, however, was little more than a steam-powered version of the long-established water-powered tilt-hammer, relying as it did on a beam to raise the hammer itself, with its own weight providing 100% of the strike force. It was 'evolution' rather than 'revolution'.

It was the Scottish engineer James Nasmyth (1807–1890), by then based in Patricroft near Manchester whose inventiveness propelled the idea into the steam age – taking it beyond anything Watt had considered.

Nasmyth was born in Edinburgh in 1808, the son of Alexander, a well respected portrait and landscape painter who has subsequently been described as the 'father of Scottish landscape art'. Alexander had worked with Thomas Telford on the aesthetics of some of his projects, most notably on the external cladding and 'castellation' of Tongland Bridge in south-west Scotland.

Although he attended Edinburgh School of Art, James did not follow in his father's footsteps, instead becoming apprenticed at the age of 21 to the engineer Henry Maudslay at his Lambeth Foundry and Engine Works where he gained a valuable grounding in engineering practice. Two years later, however, he returned to Edinburgh, briefly running his own engineering business, before re-locating to Manchester in 1832. By 1837 he had opened his Bridgewater Foundry in Patricroft near Eccles, a few miles from the city, and rapidly earned a reputation for quality.

Despite the distances involved, one of his early commercial relationships was with the Great Western Railway, for which he designed and manufactured a range of ever-larger forged tools.

So great was the demand for his products that he followed the example of Henry Maudslay and introduced a forerunner of the modern assembly-line approach to his manufacturing processes in order to increase productivity without sacrificing quality.

Once established in the Manchester area he became one of the leading lights of the Manchester Literary & Philosophical Society, and at that time the Society's members included Michael Faraday, John Benjamin Dancer, Joseph Whitworth, John Dalton, Joseph Sidebotham and many of the other leading industrialists, scientists and engineers of mid-19th century Manchester and its environs. Three members of the society were pioneer photographers – Sidebotham, Dancer and Manchester professional photographer James Mudd. All three would photograph their friend Nasmyth with his major invention, the steam hammer.

In order to achieve the quality and consistency of output from his factory which he demanded, he and his engineers devoted a considerable amount of time improving existing tools and techniques, or designing and building entirely new ones, and that was how he became involved with the idea of harnessing steam rather than just the huge weight of existing designs to improve the efficiency of stampers and tilt-hammers.

The development of the steam hammer was prompted by a letter to Nasmyth in 1838 from Brunel's marine engineer, Francis Humphries, who lamented that:

> 'I find there is not a forge-hammer in England or Scotland powerful enough
> to forge the paddle-shaft of the engine for the Great Britain! What am I to do?'

– and asking if Nasmyth's had a hammer big enough to help forge the enormous paddle-shaft for the ship which was by then already under construction in Bristol's Great Western Dock.

[above left] The steam hammer at Blists Hill Victorian Town was relocated from Thomas Walmsley & Sons' Atlas Forge in Bolton. Walmsleys had two steam hammers at the time of the works closure, and the other was relocated to a site in the grounds of the University of Bolton.

[above] By the closing years of the 19th century, the hammer had evolved into a massive and immensely powerful tool, quite different in appearance from Nasmyth's original. This photograph of a steam hammer in use at the Atlas Iron & Steel Company's forge in Sheffield was one of the illustrations in the 1897 book *The Queens Empire, A Pictorial And Descriptive Record*, published by Cassell & Co.

[right] This illustration, which appears in *James Nasmyth engineer an autobiography* written by Samuel Smiles and published in 1883, was based on a painting by Nasmyth himself. He described it as a steam hammer 'in full work'.

[below] A cross-section diagram of Nasmyth's 1851 design for his hammer.

[below right] The small steam hammer displayed at Kelham Island, Sheffield, was built by Charles Ross Ltd, at their Heeley Bridge Foundry.

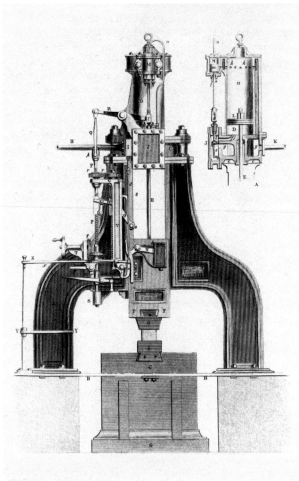

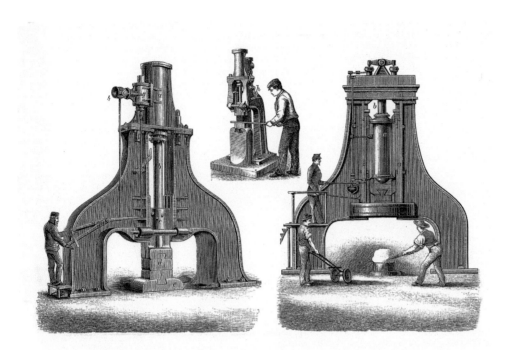

Illustrations from the German technical manual *Brockhaus' Konversations-Lexikon* published in 1894 show that the basic design of the steam hammer had changed very little in its first 50 years.

A paddle-shaft is a complex piece of engineering, and scaled up to the size required for the SS *Great Britain*, it would have required a massive foundry with a massive hammer – and that shaft would have to be built to quite tight engineering tolerances if it was to perform smoothly and reliably once in the ship.

Steam hammer technology, however, was not even in its infancy, and while Nasmyth had already made drawings of how such a tool might look and work – he sketched out his ideas in what he referred to as his 'Scheme Book' – he was still a long way from building one, let alone finding ways of producing a hammer large enough to work on a paddle-shaft of the size, weight and required accuracy for Brunel's ship.

Key to his design was abandoning Watt's beam-and-tilt design, and mounting the steam cylinder directly on top of the hammer itself. This produced a more compact, powerful and efficient design – exactly what forge-masters were looking for.

Humphries' and Guppy's chain-drive engine for the ship was already being built, and the specification for the paddle-shaft was passed to Nasmyth. Despite the fact that he had not yet built such a hammer, Nasmyth relished the challenge

According to a Mr. T. S. Rowlandson, who lectured on the history of the steam hammer to the Mechanics Institution in Patricroft on 14 December 1864 – and was described by the *Eccles Advertiser* as someone who 'has for the last twenty years filled a responsible position at the Bridgewater Foundry' – Nasmyth responded immediately to the challenge:

> 'By the same day's post, Mr. Nasmyth wrote to Mr. Humphries, enclosing a sketch of the invention by which he proposed to forge the Great Britain paddle-shaft. Mr. Humphries showed it to Mr. Brunel, the engineer in chief of the company, to Mr. Guppy, the managing director, and to others interested in the undertaking, by all of whom it was heartily approved.'

Nasmyth also gave permission for Humphries to show the drawings to 'such forgemasters as might feel disposed to erect such a hammer to execute the proposed work.' That decision would come back to bite him very quickly.

[below] This busy late 19th century industrial photograph shows a large steam hammer being used to forge part of a crank shaft. Ten men are involved in manhandling it.

[right and opposite top] From the same series of images as the large steam hammer below, two views of the completed crank shaft for an in-line four-cylinder marine steam engine. Beyond the crank can be seen a giant facing lathe which was used to mill large castings to the required fine tolerances. The location of these three photographs is believed to be one of the large forges in Wigan, Lancashire.

Long before anyone had built a big enough hammer for the task, however, Brunel had changed his mind about the ship, and made the decision to convert it to propeller drive.

The paddle shaft was abandoned, the engine was turned through 90 degrees and when the vessel was finally completed in 1845, the SS *Great Britain* was the largest screw-driven vessel yet built.

Manufacturing a large propeller shaft, while still requiring considerable accuracy, was a much simpler engineering challenge than a paddle-shaft. Humphries was reportedly distraught by Brunel's decision, Rowlandson recounting that

> 'The result was fatal to Mr. Humphries. The labour, the anxiety, and perhaps the disappointment, proved too much for him, and a brain-fever carried him off; so that neither his great paddle-shaft nor Mr. Nasmyth's Steam Hammer to forge it was any longer needed.'

In 1839, just a year after the paddle-shaft project was abandoned, a steam hammer was in operation in France at the Le Creusot Works of Schneider et Cie., its builder – the engineer François Bourdin – claiming he had invented the idea of mounting the steam cylinder

This large steam hammer, designed and patented by William Rigby, was built in 1862 by Glen and Ross of Glasgow and was one of three installed at William Parks & Company's Clarington Forge in Wigan, Lancashire.

directly above the heavy striker of the hammer. Nasmyth was able to offer the sketch in his 'Scheme Book' as irrefutable evidence that he had originated the idea in 1838, a year earlier.

After visiting Le Creusot where he was shown the hammer at work – an experience of which he would later say 'there it was, in truth, the thumping-child of my brain!' – Nasmyth instructed his advisers to register a Patent specification back home, and this they did in 1842, following up with another patent considerably improving and refining the idea in 1847. While in the 1842 patent, steam was only used to raise the hammer – its own dead weight being used on the drop – the 1847 patent introduced a powered drop, hugely increasing the power of the tool. Patent law then was notoriously loose. Until 1852 patents in England and Wales were issued under the 1624 'Statute of Monopolies', intended for a pre-industrial world. The passing of 'An Act to Amend the Law Touching Letters Patent for Inventions' in 1835, sought to formalise the process somewhat, but initially did little to add rigour to the scrutiny of patent proposals. Thus, Nasmyth was able to patent his idea despite the fact that he had not yet built actually a hammer, whereas Bourdin had.

By the time of the Great Exhibition in 1851, Nasmyth's company was producing improved steam

hammers in large numbers, for export across the world. Nasmyth's hammers were the first to be able to deliver a variable strike force, and in demonstrating this facility, he was fond of setting the hammer so it could break an egg placed in a wineglass without breaking the glass, while then using the same hammer to deliver a strike of such force that it could shake the entire building. Such demonstrations of controlled power drew large audiences.

Others would patent improvements to the simple hammer, automating its functions and increasing the number of strikes per minute it could deliver. One interesting example, a double-acting 20cwt hammer designed and patented by Robert Morrison of Ouse Burn Engine Works in Newcastle-on-Tyne, was exhibited at the 1862 International Exhibition in London and drew considerable crowds when it was used to crack a nut – a variation on Nasmyth's egg and wine glass trick.

In its various forms – and many different sizes – the steam hammer became one of the most important and ubiquitous industrial tools of the nineteenth century, dramatically cutting production time and costs across a wide range of heavy industries.

By the time Brunel was building his SS *Great Eastern* in the late 1850s, which would have both propeller and paddle drive, the steam hammer was an essential tool in its construction, used in the forging of the its enormous paddle-shaft.

Before then, however, having made a considerable fortune, Nasmyth had been able to retire in 1856 at the age of just 48 – the year before the *Great Eastern* was launched – and enthusiastically devoted his life to his fascination with astronomy.

He later became a good friend of James Carpenter, an astronomer at the Royal Observatory in Greenwich – who shared his ill-founded conviction that the topography of the moon had been created by volcanic action – and the two men embarked on a major project to develop their theories and publish the results in a profusely illustrated book *The Moon: Considered as a Planet, a World, and a Satellite* published by John Murray in 1874.

It was illustrated with images based on drawings, paintings, and plaster models which Nasmyth had made and then photographed. The Nasmyth Crater and the Carpenter Crater – two impact craters on the moon – are named in their joint honour.

Their theories may have been superseded, but the importance of their work, and the accuracy of Nasmyth's modelling on which the illustrations were based, made their book an important and fascinating contribution to science.

His 'autobiography' – *James Nasmyth, Engineer* – published in 1883 by John Murray was edited, and in all probability largely written, by Samuel Smiles, who assembled it from Nasmyth's own uncompleted manuscript. In it, Smiles wrote

'the most engrossing of Mr. Nasmyth's later pursuits has been the science of astronomy, in which, by bringing a fresh, original mind to the observation of celestial phenomena, he has succeeded in making

Morrison's Patent Double-Acting Steam Hammer drawing a crowd at the 1862 International Exhibition in London – an illustration from *The Penny Illustrated*, Saturday, 27 September, 1862. In this interesting design – which offered the operator considerable control over both strike frequency and strike force – the piston and the hammer shaft were forged out of a single piece of metal.

some of the most remarkable discoveries of our time. Astronomy was one of his favourite pursuits at Patricroft, and on his retirement became his serious study. By repeated observations with a powerful reflecting telescope of his own construction, he succeeded in making a very careful and minute painting of the craters, cracks, mountains, and valleys in the moon's surface, for which a Council Medal was awarded him at the Great Exhibition of 1851.'

He also turned his pair of telescopes towards the sun, sketching and painting what he saw, prompting Smiles to observe that

'Some two hundred years since, a member of the Nasmyth family, Jean Nasmyth of Hamilton, was burnt for a witch – one of the last martyrs to ignorance and superstition in Scotland – because she read her Bible with two pairs of spectacles.

Had Mr. Nasmyth himself lived then, he might, with his two telescopes of his own making, which bring the sun and moon into his chamber for him to examine and paint, have been taken for a sorcerer.

But fortunately for him, and still more so for us, Mr. Nasmyth stands before the public of this age as not only one of its ablest mechanics, but as one of the most accomplished and original of scientific observers.'

[above left] Nasmyth with one of his home-made telescopes, from *James Nasmyth Engineer; an Autobiography* by Samuel Smiles.

[above] Plate XV 'Mercator & Campanus' from the book, *The Moon: Considered as a Planet, a World, and a Satellite*.

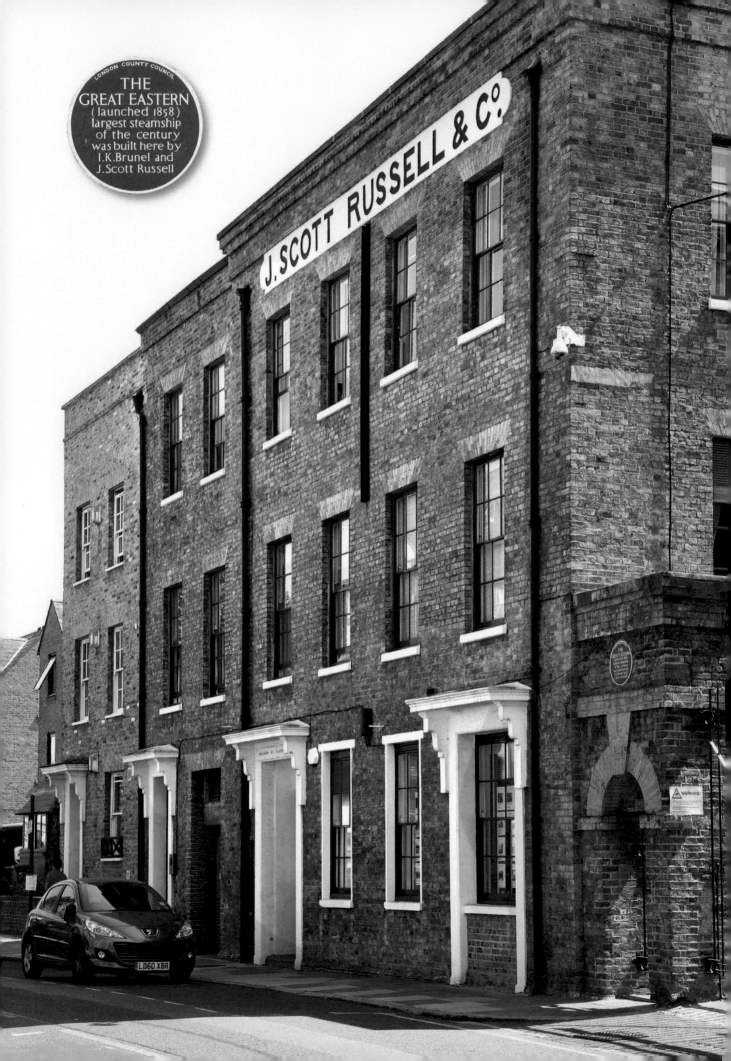

JOHN SCOTT RUSSELL – MAKING WAVES

The shape of ships' hulls owes a great deal to the pioneering work of John Scott Russell (1808–1882), one of Victorian engineering's most innovative figures, yet one whose achievements have rarely been properly celebrated.

Although he is best remembered as Isambard Kingdom Brunel's partner and co-designer of the largest steamship built in the 19th century, the SS *Great Eastern*, and took more than his share of the criticism when the ship's construction and launch did not go according to plan, Russell has for too long stood in Brunel's shadow as far as published accounts of 19th century shipbuilding go. Indeed some writers have almost dismissed the significance of his role in the design and construction of the vessel – a role Brunel also tried to diminish.

But shipbuilding was only one part of his pantheon of engineering achievements. Russell was responsible for many other pioneering developments during his long career and his contribution to Victorian engineering on both land and sea should not be vested in the memory of a single ship – he deserves much wider recognition.

[above] A contemporary portrait of John Scott Russell.

[below left] A contemporary engraving of the SS *Great Eastern* during the ill-fated first attempt to launch her.

[opposite page] J. Scott Russell & Company's building on Westferry Road, Millwall now carries a blue plaque to commemorate the building of the SS *Great Eastern*.

[inset] The blue plaque over the archway at the former J. Scott Russell & Co. building in Millwall.

connections with:

- James Nasmyth
- William Fairbairn
- Isambard Kingdom Brunel
- George Stephenson

[right] The 1850s Plate House building, where the plating for the SS *Great Eastern* was forged, can be seen in the background of many of the photographs taken during the construction of the great ship.

[below] The SS *Great Eastern* under construction, from the *Illustrated London News*, 30 May 1857. The picture is captioned 'The Great Eastern Steamship on the stocks, Millwall, 22,500 tons bur[d]en, from a photograph in the possession of Mr. Scott Russell.' The original photograph would probably have been taken by Brunel's preferred photographer Robert Howlett.

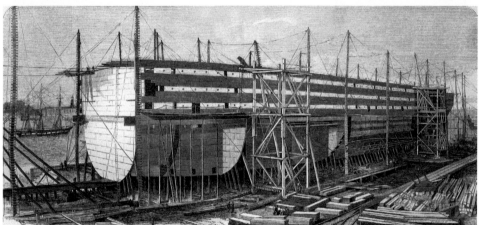

He applied a scientific mind and engineering precision to all his projects, and on 16 January 1860 was central to the foundation of The Institution of Naval Architects, now the Royal Institute of Naval Architects.

At their first meeting the founders declared 'We, who are present, do now constitute ourselves an Institution of Naval Architects for the purpose of advancing the science and art of Naval Architecture'. For the 52-year old engineer, that was just the latest achievement in an already long and eminent career.

Russell had been born in 1808 in the village of Parkhead – then some miles from Glasgow. His father, David, was a recently qualified preacher, teaching in the local parish school, a role he fulfilled while waiting a suitable appointment as a minister in the 'Relief Church', one of several groups which had broken away from the Church of Scotland in the second half of the 18th century and the first half of the 19th. John's mother had died not long after his birth and, by the time he was four years old, his father had remarried and moved the family to Hawick, later taking up another appointment in Errol near Dundee at the princely salary of £85 per year.

John Russell was only 13 years old when he was admitted to Glasgow University – where years before, James Watt had worked as an instrument technician – and it was around that time that he added 'Scott' to his name, a tribute to his mother. It was as 'J. Scott Russell' that he later established his shipyard on the Thames.

Long before he began to make a name for himself in the world of shipbuilding, he had developed an interest in steam traction on land. The idea of replacing horses by steam engines in road vehicles had occupied several budding engineers – including 20-year old James Nasmyth whose short-lived steam carriage had operated on the Queensferry Road on the outskirts of Edinburgh in 1827. Russell, unlike several of the others who launched steam carriage services in London and Gloucester in the early 1830s, quickly identified passenger comfort as a key to success. His 1834 steam coach was well sprung and apparently had a beautifully upholstered interior. At the rear was a simple two-cylinder vertical steam engine with a 12ins stroke.

He patented his design and sold the idea to the Steam Carriage Company of Scotland, who commissioned the building of six coaches from Grove House Engine Works of Edinburgh, and these entered service in late March between Glasgow and Paisley. Some sources claim the passenger capacity was 26, but contemporary illustrations of the vehicle make that seem unlikely. Coal and water for the engine were carried in a trailer towed behind the coach. *The Glasgow Herald* newspaper carried a report on the service about four weeks after its inauguration.

> 'On Tuesday last a single carriage belonging to the Steam Carriage Company of Scotland performed the most successful runs that have ever been accomplished on common roads, having gone six successive trips with passengers between Glasgow and Paisley, and in an average of forty-three minutes... ...being a distance in all of forty-six miles in four and a half hours– at a rate of more than 10 miles an hour.'

The company's plan had originally been to run between Edinburgh and Glasgow, but the owners of the toll road refused permission, claiming that the steam carriage would destroy the road surface and suggested imposing punitive toll charges to defray the anticipated repair expenses. The service lasted less than a year, the carriages being sold at auction in London in February 1835, but by that time, Russell had already turned his attention to the phenomenon of the bow wave created by a boat as it cut through the water, initially on canals but later also on the open sea. Recognising that pushing a large quantity of water ahead of it was not an efficient use of either horse-power or steam-engine power, he embarked on a series of key experiments. The first boat he built to explore the phenomenon in the controlled environment of the Edinburgh and Glasgow Union Canal he, appropriately, named *Wave*.

The results of his experiments led to what he called his 'wave-line hull' – the shape of which displaced the least possible amount of water necessary to keep the bow wave at a minimum. His resulting 1836 paper 'The Ratio of the Resistance of Fluids to the Velocity of Waves' was

Russell's innovative steam carriage from 1834 was designed for passenger comfort.

145

groundbreaking in its day. Just 20 years later, the principles of his 'wave-line hull' would inform the design of the SS *Great Eastern*.

In the *Report of the Fourteenth Meeting of the British Association for the Advancement of Science* which took place in 1844 and was published in 1845, two of Russell's papers were included exploring his observation on the generation and impact of waves in canals, in rivers and at sea.

His work on the impact of waves had been initiated by a logical progression from his steam carriage of 1834 – he planned to build a steam-powered passenger vessel to operate on the canal between Edinburgh and Glasgow. He quickly became aware of the fact that, while of little consequence with a small canal boat being hauled slowly by a horse, a larger steamboat with its blunt prow was pushing a larger and larger quantity of water ahead of it as it picked up speed.

In the confines of a narrow canal – especially a canal built to James Brindley's specifications – that water could not be easily dispersed to the sides, so had nowhere to go except forwards. A significant amount of energy was thus being expended to push the water when it should more efficiently be being used to propel the vessel.

The design of ships' hulls had changed little in centuries, largely dictated by both long-established tradition and the limitations imposed by the wide hull design needed to ensure the stability in rolling seas of wooden ships with tall masts.

Steam-powered vessels were nothing new – William Symington's steam barge *Charlotte Dundas* had been built to operate on canals as early as 1803, but the experiment had been abandoned amid fears that the bow wave and wash were damaging the banks.

Henry Bell's small paddle steamer *Comet* had carried its first paying passengers in 1812, but that was on the River Clyde. To a world which had never previously seen a steamer, these new craft, belching smoke and without sails, were initially a source of surprise and sometimes even fear. But while canal owners sought to ban steamboats, Russell set out to apply scientific principles to finding a solution to the bow wave problem. That answer lay in combining two approaches – the mathematical and the empirical – calculation and experiment. To that end he built a scale model of a canal in his back garden. His 'test tank' – the first in the world – was thirty feet long and he used it in 1834 to experiment with how his 'Great Wave of Translation' built up, and how it dispersed. Having made endless notes and calculations based on the wave tank experiments, he then observed how full size vessels behaved in canals, rivers, and at sea. In his British Association paper he explained his thinking:

> 'I was observing the motion of a boat which was rapidly drawn along a narrow channel by a pair of horses, when the boat suddenly stopped—not so the mass of water in the channel which it had put in motion; it accumulated round the prow of the vessel in a state of violent agitation, then suddenly leaving it behind, rolled forward with great velocity, assuming the form of a large solitary elevation, a rounded, smooth and well-defined heap of water, which continued its course along the channel apparently without change of form or diminution of speed. I

Detail from an Edwardian postcard showing the bow wave created by a slow-moving horse-drawn barge. This view was photographed on the relatively wide Leeds and Liverpool Canal. On narrow canals the bow wave built up in front of the boat.

followed it on horseback, and overtook it still rolling on at a rate of some eight or nine miles an hour, preserving its original figure some thirty feet long and a foot to a foot and a half in height. Its height gradually diminished, and after a chase of one or two miles I lost it in the windings of the channel. Such, in the month of August 1834, was my first chance interview with that singular and beautiful phenomenon which I have called the Wave of Translation'

He first observed the behaviour of his wave by building a long tank with a baffle at one end holding water behind it at a higher level, and then measuring and recording what happened when the baffle was removed and the 'excess of water' started to make its way along the tank.

He then introduced model hulls with very different characteristics into the tank and observed how they made their way through the water, recording everything that happened both to the wave they formed and the water they displaced.

The *Charlotte Dundas* as depicted on a Wills (top) and Churchman's (above) cigarette card published in 1911 and 1936 respectively. The bow wave and wash were cited as reasons why steamboats should not be used on canals.

The ideal hull shape, Russell argued, 'should be such as to remove the articles of water out of the way of the ship just sufficiently far to let the largest section pass, and not a jot farther.' In theory he was absolutely right, but putting that theory into practice was much more challenging – the wide-bodied hulls of traditional sailing vessels, and the even greater obstacles created by the paddle boxes of steamers did not easily submit to the theoretical ideals.

History has consistently undervalued Russell's discoveries, and it was not until the 1960s that computer modelling confirmed the significance of his proposals.

Russell's interests were wide-ranging, and few were surprised when, having already contributed entries on steam engines and steam navigation for the 1841 edition of the *Encyclopaedia Britannica*, he moved away from the world of shipbuilding and in 1844 took on the editorship of *The Railway Chronicle*, while also writing about railways for Charles Dickens' new newspaper, *The Daily News*.

His many other interests included writing an important paper on the mathematical imperatives in suspension bridge design if wind-induced vibrations were to be avoided.

As Secretary of the Royal Society of Arts in the late 1840s, he was one of the originators of the idea of a Great Exhibition, and instrumental, with Henry Cole, in getting Prince Albert involved in the project. His organising skills would be recognised by the committee charged with overseeing plans for the 1851 Crystal Palace exhibition in Hyde Park, and the Prince appointed him as its secretary. Despite its royal patronage, the Government initially declined to provide support for the project – offering 'not even the red tape to tie up the papers'. When the exhibition turned out to be both a great success and highly profitable, the Prime Minister Lord John Russell – no relation – must have been relieved that they had relented, otherwise they would have had not been able to claim any share of the glory.

Russell's involvement with the Crystal Palace did not end when the gates were closed on the exhibition for the last time. He became a partner in the new Crystal Palace Company which was established to acquire Joseph Paxton's Hyde Park exhibition building and relocate it to the new Crystal Palace site at Sydenham.

Before then, however, he had turned his attention back to shipbuilding, and become a partner in resurrecting the fortunes of William Fairbairn's failed Thames-side shipyard on the Isle of Dogs at Millwall. The consortium initially leased the yard, but Russell later bought it.

Shortly after Russell took over at Millwall, Brunel approached him to tender for the construction of two iron steamships. They would be the SS *Victoria* and the SS *Adelaide*, built for the Australian Royal Mail Steam Navigation Company and the first vessels on which Brunel and Russell would collaborate. After those, their next project together would be the SS *Great Eastern*. While Brunel had overall responsibility for the design of the *Adelaide* and the *Victoria*, Russell made significant contribution to the hull design, including the use of both transverse and longitudinal strengthening bulkheads, essential in giving a large iron ship structural stability at sea. Both vessels also incorporated one of Russell's most innovative features – iron decks, which greatly increased the structural rigidity.

However when the 680ft long SS *Great Eastern* was being planned, both Russell and Brunel realised that extra stiffening would be needed, otherwise it risked breaking its back if it crested rather than cut through the waves.

Informed by George Stephenson and William Fairbairn's box girder construction for the Britannia Railway Bridge across the Menai Strait, they evolved the double-skinned hull design for the *Great Eastern* using iron plates riveted either side of a framework of iron ribs, creating rigid 'cells'.

[below] The Machine Hall at the Great Exhibition in the Crystal Palace in Hyde Park, 1851.

[bottom] The interior of one of the Crystal Palace transepts, engraved from a daguerreotype by the eminent photographer J. J. E. Mayall.

Russell's original tender for building the ship – including the paddles and their engines, but excluding the engine for the screw propeller, interior fixtures and fittings, and launch costs – was a little over £250,000. The contract for the propeller engine went to James Watt & Company, successor to Boulton & Watt.

It went hugely over budget, of course, and tested Russell's friendship with Brunel to the limit on several occasions. Brunel, it appears, was less than happy to see Russell gain public credit for his input into the progress of the ship, and quite quick to blame him when things went behind schedule. Before the hull was even completed, they were on the brink of falling out irrevocably – and disagreement about how the ship should be launched only made matters worse.

The outcome was inevitable – Brunel demanding changes for which he felt no obligation to pay, withholding payments due to Russell – led to delays and the ultimate bankruptcy of the Scott Russell Shipbuilding & Engineering Company.

By May 1856, Russell was back supervising the completion of the ship – but as an employee of the Great Eastern Steam Navigation Company under Brunel's sometimes fickle and contradictory instructions. It was a short-lived arrangement and he was soon relieved of his position for allegedly stepping beyond his newly restricted authority. As the vessel was being built next door to Russell's yard, however, his name remained closely associated with it.

Despite the ground-breaking nature of Russell's work in the 1830s and 40s, it was one of his contemporaries working in the same field of hydrodynamics – William Froude (1810–1879), an engineer and naval architect from Devon – to whom history has given much of the credit. Froude was already working on the Bristol to Exeter section of Brunel's Great Western Railway, and it would be he who would get the credit for establishing the formulae for efficient hull design. His first test tank, a modest affair, was built in the attic of his house in Torquay in 1870, more than a quarter of a century after the 30ft tank in Russell's garden was first used. Perhaps not surprisingly, given Brunel's antagonism towards Russell after the launch of the SS *Great Eastern*, it was he – who had already used Russell's design principles in at least three of his ships – who directed Froude towards the subject. Froude's work developed some of Russell's ideas and gave them a sound practical as well as mathematical foundation.

Especially important was his realisation that the relationship between the shape and length of a ship's hull and the vessel's design speed through the water was critical. The ideal hull shape for a vessel moving slowly along a canal would be very different to that or a large vessel moving at a much greater speed. He measured the bow wave and stern wave – and thus the resistance – of each model hull as it was towed at a given speed along his test tank, and thus generated a 'Froude

Number' for each given set of circumstances. His attic test tank quickly confirmed the value of practical experimentation, and the Admiralty commissioned a much larger tank which was brought into service at Chelston Cross in Torquay in May 1872. There he could test model hulls up to twelve feet in length, moving them at a scale percentage of the projected speed of the full-size ship.

The use of a test tank by naval shipbuilding yards soon spread worldwide, and it is reckoned upwards of three hundred such facilities were built. Today, much of the work for which they were used is now undertaken using computer modelling, but still informed by Russell's ideas and principles, and Froude's calculations.

By 1858, Russell's standing as a shipbuilder had been restored, and his advice had been sought by the Admiralty over the planned construction of HMS *Warrior* which would be built

[above] Russell's elegant wave-line hull taking shape as the vessel – then referred to as the *Leviathan* – dominated the Thames shoreline as it was being constructed.

[below] Russell's contract for the SS *Great Eastern* included building the giant paddle wheels and engine, which had to be installed into the hull before the ship was launched. The engines, illustrated in *The Illustrated London News* on 3 September 1859 worked so well that the eminent American engineer Alexander Lyman Holley, declared 'Four great oscillating cylinders on one shaft, and you would not know they were in motion save for the light rush of steam at their centres. There is no pounding of boxes and valves, or general rattling jar, thump and spring, as is too common with oscillating engines.'

at the Thames Iron Works' yard. He had been advocating ironclad warships since the early 1850s and his design for *Warrior* would be immensely strong – with both longitudinal and transverse watertight bulkheads – and with the hull backed by thick timbers to stop the iron plates shattering in the case of a direct hit.

Of course, Russell's hull design helped the warship achieve her high speed, and the concentration of her heaviest armour amidships gave her good stability and maneuverability. However he got only minimal credit for his work on the game-changing vessel which now sits, restored to her former glory, in Portsmouth's Historic Dockyard, more than 160 years after she was launched.

That his design was, in effect, compromised to some extent by the strict requirements of his Admiralty overlords was a source of some annoyance to him – something which he vented while leading a discussion following on from Edward James Reed's paper 'On Long and Short Iron-clads' delivered to the Institution of Naval Architects in March 1869. Reed was the Chief Constructor of the Navy, at the time and the Institution's Vice President, while Russell was President. He asserted his claim as the designer of *Warrior* with the words

'I can venture, perhaps, to speak with some authority upon that subject, for it is, I think, understood that the Warrior class was invented and created jointly between the Controller's Department and myself. We have agreed to say that we are joint and equal owners in the credit and discredit which may attach to the Warrior class—I mean, as technical men.'

[above] The test tank at William Denny & Brothers yard in Dumbarton – now one of the Scottish Maritime Museum's two sites. The tank is 300ft (91m) long and as the test hull is moved along its length at controlled speeds, its bow waves could be analysed. Such analytical research in ascertaining optimum hull design is now done using computer modelling. The museum has also preserved a hull-shaping machine based on William Froude's original invention.

[left] Russell and Froude would see much that they recognised in the bows of Britain's *Queen Elizabeth Class* aircraft carriers.

[opposite page] The ironclad HMS *Warrior* launched in 1860 – the first of the Warrior-class warships – was designed by John Scott Russell and naval architect E.J.Reed, embodying Russell's hull designs principles.

WILLIAM ARMSTRONG, AN INDUSTRIAL TITAN

The link between the pioneer of electric lighting, Joseph Wilson Swan, and Sir William George Armstrong (1810–1900), one of the most important industrialists of the Victorian era, initiated the idea of light being available at the flick of a switch. Swan had gone into partnership in 1846 with his brother-in-law John Mawson, a pharmacist originally from Penrith in Cumbria, to create the Newcastle company of Mawson & Swan Ltd, photographic chemists, whose collodion – the potentially lethal mixture of guncotton and ether which was used in early photographic plates – was said to be the finest on the market.

Mawson went on to become Newcastle's sheriff, a post he held until his untimely death in an explosion in 1876 while trying to dispose of some nitro-glycerine, but his work with Swan in the field of photographic chemistry, and in particular the perfecting of collodion dry plates, highlights him as an important pioneer.

Mawson & Swan manufactured and marketed a range of photographic chemicals and equipment – work carried on by Swan after Mawson's death – but Swan is best remembered as one of the fathers of the electric light, and he demonstrated his first incandescent filament light bulb in 1878. His work was noticed by Armstrong, who had established his engineering works on the River Tyne at Elswick in 1847.

[above] A portrait of Armstrong, taken c.1877 when he was 67 years old.

[below left] Sir William Armstrong's original fishing lodge near Rothbury grew into a magnificent and highly modern mansion.

[below right] The National Trust's recreation of a Swan Lamp in the Library at Cragside. The originals were the first incandescent electric lights to be installed in a domestic property.

[opposite] The Engine House at Ellesmere Port still houses two of the four Armstrong engines which powered the docks' hydraulics.

[top inset] The maker's plate on one of the engines.

[bottom inset] The 36.5hp duplex cross-coupled engines, were built at Elswick in 1873.

connections with:

• William Jessop
• Thomas Telford
• Isambard Kingdom Brunel
• William Arrol

Despite his later successful career in heavy industry, Armstrong initially trained as a solicitor, and practised law from 1833 until 1847 when he set up the Elswick Works. Long before that, however, his engineering curiosity had taken hold.

By the late 1830s he had built what he referred to as his improved rotary water motor, and his fascination with the power of water, and his curiosity as to how that power might more efficiently be harnessed, led to a growing interest in the science and engineering of hydraulics. In the early 1840s he published several papers proposing that the pressure of water in mains pipes could be used to work cranes and other machinery.

Within a remarkably short period of time, he had developed a hydraulic crane and, as his understanding expanded, the company would develop ever more sophisticated hydraulics. Over the years the Elswick Works developed into a powerhouse of innovation, and brought Armstrong considerable wealth.

In the 1850s, as befitting a man approaching middle age and of great wealth, he decided to build himself a country house, a fishing lodge, near Rothbury in Northumberland and, completed in 1863. What a house it turned out to be.

Cragside, set in a wooded landscape, grew over the years to become a magnificent mansion, but it was Armstrong's fascination with Joseph Swan's experiments which led, in 1879, to it becoming the first house in the world to be illuminated by Swan's incandescent lamps, the system powered by the world's first hydro-electric scheme – designed by Armstrong himself, of course.

Parts of the house had been fitted with carbon arc lamps as early as 1878, powered by a small turbine which was fed by waters from Debden Lake. Power was generated by a Siemens dynamo and then relayed to the house along nearly a mile of cable. When the arc lights were not needed, the power was diverted to operate the estate's sawmill. The major step forward, however, took place in 1879 and saw the first of Swan's lamps installed.

The first room to be illuminated by Swan lamps was the library, and Armstrong wrote to the journal *The Engineer* about the system in January 1881 – by which time there were over 45 such lamps in use in the house. He described the eight lamps in the library – four of them

[right] The Water Wheel outside the Power House was originally used to generate hydraulic power on the estate.

[far righ]: What was, at the time, 'state of the art switch-gear' was used to control the electrical system, all under the supervision of the 'Caretaker of the Electric Light'. On the floor are replacement blades for the turbines which were at the heart of the system.

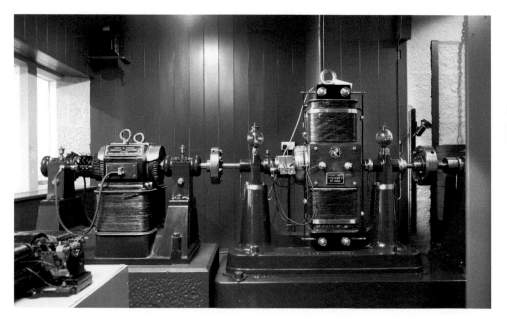

[left] A small Siemens dynamo sits in front of the Crompton generator which is connected to a Thomson Double Vortex turbine which in turn was fed by water at 150psi through a 7-inch pipe from the lake above the power house.

[above] The 'Pelton' Vortex water turbine by Gilbert Gilkes & Gordon was installed in 1883, replacing an earlier and less powerful model. Gilkes turbines came in a range of sizes.

in a large globe suspended from the ceiling, and four others each built into ornate enamelled vases – writing:

> 'The vases, being enamel on copper, are themselves conductors, and serve for carrying the return current from the incandescent carbon to a metallic case in connection with the main return wire. The entering current is brought by a branch wire to a small insulated mercury cup in the centre of the base, and is carried forward to the lamp by a piece of insulated wire which passes through the interior of the lamp on the top. The protruding end of this wire is naked, and dips into the mercury cup when the vase is set down. Thus the lamp may be extinguished and relighted at pleasure merely by removing the vase from its seat or setting it down again.'

Armstrong embraced electricity enthusiastically, replacing the turbines and dynamos when the demand for electricity required a bigger output – all these improvements taking place in a bigger powerhouse he had built at Burnfoot in 1886 and fitted out with a water turbine by Gilbert Gilkes & Gordon, and a Crompton generator. The new powerhouse was fed with water from Nelly's Moss Lakes over 300ft feet above it, and could generate almost 18kW of power. The generator, designed in 1883, is still in situ today. All this was supervised 24-hours a day by the splendidly titled 'Caretaker of the Electric Light'.

The dynamo was capable of generating 90 amps of 110 volt direct current, but even that soon proved insufficient to meet peak demand from the house. So, in the mid-1890s, a battery room and a second dynamo room were added to the power house, the former filled with an array of huge lead-acid batteries, each weighing nearly one hundredweight. They were charged at times of reduced demand, and then drawn upon to meet peaks, or when the water flow was insufficient to operate the dynamo at full capacity.

The Caretaker of the Electric Light was able to communicate with the house by telephone and also with any other agency he might need to contact in the course of his work. By the mid-1890s, that included the Rothbury Gas Works which Armstrong had also built. With a gas engine installed in Cragside's power house to generate more electricity whenever needed, the drain on the town's gas works might have impacted on all its other users, so the Caretaker

[above] One of the 100-ton guns being manufactured at Elswick in 1879. The Elswick Ordnance Company built 15 of these massive guns. Four each were installed on two Italian battleships, the *Caio Duilio* and *Enrico Dandolo* while others were used to strengthen fortifications on Malta and Gibraltar in case those same battleships were used against British interests. The Malta and Gibraltar guns still survive.

[below] On the gun-decks of HMS *Warrior,* a replica of Armstrong's 110-pounder seven-inch breech-loading, rifled barrel guns, introduced in 1859. The gun had a range of two and a half miles, huge for its day.

of the Electric Light was required to telephone the gas works before starting the engine.

Despite his enthusiasm for both electricity and hydraulics, Armstrong first established his reputation in the field of military ordnance during the Crimean War, 1854–56, developing and manufacturing a lighter and much more versatile field gun for the British Army. Heavier guns followed, and by the end of the decade Armstrong guns were the artillery weapons of choice. They employed an innovative method of construction of a wrought iron tube over which were shrunk a number of wrought iron sleeves, which added strength.

The new designs were formally adopted by the British authorities in 1859, manufactured by the Elswick Ordnance Company, supervised by Armstrong, but separate from his other enterprises as he had taken on an official role as Engineer of Rifled Ordnance at the Royal Gun Factory. He ceded his patent to the Elswick Ordnance Company and was knighted by Queen Victoria shortly afterwards.

These patented guns were the first 'modern' guns, breaking away from the cannon which had been typical ordnance since mediaeval times. Instead of a round muzzle-loaded cannon ball, breech-loaded tapered shells could be used, hugely increasing the range and accuracy of the weapons.

There were initial problems with the breech-loading system and the rifled barrels – causing the guns to and overheat and jam, requiring careful cleaning, prompting the government to abandon the guns in 1863 and returned to muzzle-loading. It took 17 years before Armstrong convinced the authorities that the problems with the system had been eliminated, and breech-loaders were re-introduced in 1880.

Since 1863 the inner gun barrel had been made of toughened mild steel and later barrels were wire-wound using steel wire, giving them much greater strength. That assembly method endured for many years and was used in the manufacture of 100-ton guns, then the largest in the world.

After his breech-loading system was once again deemed acceptable by the authorities, Armstrong designed and built the 12ins Mark VIII naval gun, one of the first large British

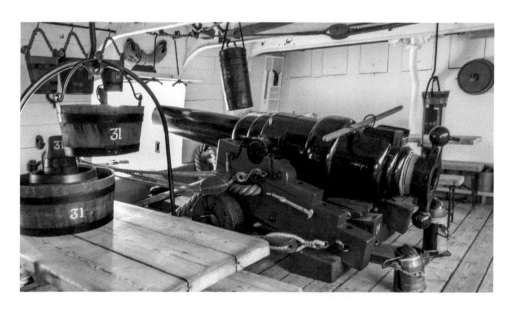

rifled breech-loading naval guns. It was designed to cope with the higher pressures generated by the new cordite propellant of the 1890s.

Although the company had been making smaller gauge wire-wound guns for some years, this was the first large wire-wound weapon and represented a major development both for Armstrong and for the Royal Navy.

One of the first vessels to be equipped with the new Armstrong Mark VIII breech-loading wire-wound guns was HMS *Magnificent*, launched in Chatham in 1894.

Sir W. G. Armstrong, Mitchell & Company – as the company was known after a merger with the Mitchell shipyard in 1881 – became a major constructor of warships at the Elswick yard, while merchant ships were built at Mitchell's.

A further merger with Joseph Whitworth & Company of Manchester in 1897 led to another name change, this time to Sir W. G. Armstrong, Whitworth & Company.

In 1846 – the year before he established his engineering base at Elswick – Armstrong had set up the Newcastle Cranage Company to develop his ideas for hydraulic dockside cranes, which he had first proposed more than a decade earlier, perhaps aware of the achievements of Joseph Bramah and Henry Maudslay.

By the time of the Great Exhibition in 1851, with his workforce already exceeding 400 people, Armstrong hydraulic cranes were in widespread use and the example displayed at the Crystal Palace was widely praised.

Jesse Hartley, the naval architect and prime mover behind Liverpool's Albert Dock, actually travelled to Elswick in 1847 to see Armstrong's cranes in action, immediately ordering two of them at £500 each, together with some hydraulic hoists. Liverpool Corporation's mains water

[above] The rifled barrel of an Armstrong 110 pounder gun on HMS *Warrior*.

[below left] Field-gun assembly at Elswick c.1905.

[below] An 1887 illustration of Armstrong's 'Elephant Sectional Gun'.

[bottom] The artillery camp at Balaclava, one of Roger Fenton's series of photographs taken in spring 1855 at the height of the Crimean War.

had sufficient pressure to power them, and Hartley even persuaded the council to meet the cost of laying the pipes.

While Armstrong's original idea had been to use mains water to operate all the hydraulics, he quickly realised that there were many places where there was insufficient water or insufficient pressure to power his cranes, and for such locations he needed a means of creating that pressure so it could be called upon when needed.

The hydraulic accumulator proved to be the answer – a tall tower in which water could be stored until needed, then released to perform its work.

Hartley was clearly an advocate of Armstrong's hydraulic equipment, and a later dock engineer at Liverpool, Anthony George Lyster, actually underwent some of his training at the Elswick Works before returning to Liverpool to work on the expansion of the docks. That took place in the 1870s, at a time when increasing traffic in the docks was putting considerable demands on Liverpool Corporation's water supply.

The answer was to build an accumulator tower and pumphouse, and equip it with two powerful steam engines to feed Armstrong's cranes, winches and dock gates.

One of the most recognisable water accumulator towers – containing 33,000 gallons of water – was built between 1849 and 1852 at Grimsby Docks, but the system was already obsolete by the time work on it finished, as Armstrong had developed his improved weighted accumulator in which steam power was used to raise the water up into the accumulators and maintain it there under considerable pressure until required.

A number of notable examples of Armstrong's hydraulic system survive including two pumping engines and weighted hydraulic accumulator tower at Ellesmere Port docks. The tower contains a 17ins diameter weighted hydraulic ram with a stroke of 12ft and it carries a load of around 70 tons of pig-iron housed in a cylindrical container. This provided a pressure

[clockwise from below] The cruiser *Yangwei* under construction for the Chinese Navy at Elswick in 1881. Elswick also built ships for the Japanese, Italian, Chilean, Brazilian, American, Spanish and Ottoman navies.

The 1901 Chatham-built Duncan Class battleship HMS *Albemarle*, was equipped with Armstrong's 12ins guns.

The giant rotating gun turrets for Dreadnought-class battleships under construction at Armstrong's factory c.1905. These massive 12ins guns became the standard heavy armament on Britain's most powerful battleships. The guns and their heavily-armoured turret weighed more than 46 tons.

of 750lbs per square inch pressure. Aspects of the hydraulic system at Underfall Yard in Bristol's 'floating harbour' can still be seen.

One of Armstrong's most impressive systems is that which once powered London's Tower Bridge – with horizontal twin-tandem compound steam engines and their accumulators, built in 1894. The bridge was considered an engineering marvel, powered by the latest in Victorian machinery. It was the largest bascule bridge ever built. The engines at the heart of it were rated at 150hp each, with 18ins diameter high pressure cylinders and 30ins diameter low pressure cylinders. The giant nine-feet diameter flywheel on each engine was cast as a single piece – a remarkable feat of engineering craftsmanship.

Steam was provided by four Lancashire boilers built for Armstrong by the Crosthwaite Furnace Company of Leeds and, like the engines themselves, these were installed in a barrel-vaulted engine house beneath the roadway on the south side of the river. The engines operated without major mishap for 80 years until replaced by electric motors in 1974.

A third engine – also by Armstrong, Mitchell & Company – was added in 1941 (Engine No.1190SE) but this was removed in 1974 when electricity replaced steam power, and is now on display in the Forncett Steam Museum in Norfolk.

During the first month of operation in 1894, the bridge was opened and closed on no fewer than 650 occasions – more than 20 times a day, so busy was traffic on the river. Back then, opening was 'on demand', but today the bridge authorities require 24 hours' notice before an opening can be scheduled.

The restored engine house was opened to the public in 1982, but since then, like the bridge itself, the engines have been turned by electricity rather than steam.

The engines did not actually power the bridge's lifting mechanism. Their job was to pump water into six huge 100ton vertical accumulators – two in the south bank engine house and two in each pier – where it was also held under 750lbs per square inch pressure, the same as at Ellesmere Port.

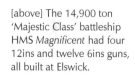

[above] The 14,900 ton 'Majestic Class' battleship HMS *Magnificent* had four 12ins and twelve 6ins guns, all built at Elswick.

[below] From the time Jesse Hartley's Royal Albert Dock in Liverpool was constructed in the 1840s, all the dockside cranes and internal hoists were driven by Armstrong's hydraulics. The enclosed fireproof warehousing was inspired by Thomas Telford's St. Katharine Dock in London.

The energy stored in the accumulators was used to drive the eight three-cylinder, single-acting, hydraulic engines which actually raised and lowered the bridge's 1,200ton counter-balanced bascules. Depressing the plunger in one of the accumulators generated enough power to open and close the bridge twice. One of the original hydraulic lifting engines is on display. Since the steam engines were retired, oil under pressure powered by electro-hydraulic motors has been used to operate the lifting mechanism.

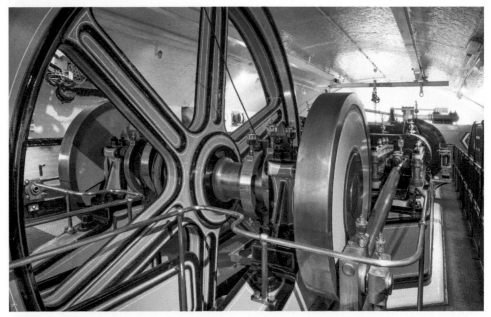

[top] Tower Bridge's boilers used water from the River Thames. The Hotchkiss Circulators removed mud, air and oil from the system.

[above] Six accumulators were used to store water under pressure to drive the hydraulic bascule engines.

[above right] One of the 1893/4 Armstrong, Mitchell & Co. engines at London's Tower Bridge, now restored and turned by electricity.

[right] The 1894 hydraulic bascule engines, withdrawn from service in the 1970s and now on display in the engine room.

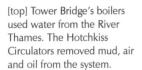

Another of Armstrong's great hydraulic projects, undertaken at the same time as the Ellesmere Port contract, was the construction of Newcastle's famous Swing Bridge over the Tyne, built at a cost of around £240,000 and opened to traffic in 1876. When first opened, it was the largest swing bridge in the world.

The 1,300ton rotating section can be opened and closed in 6 minutes, but early on in its life, there were complaints that it could take up to 45 minutes to raise steam before the bridge could be operated – a curious complaint as once the accumulators were charged with compressed water – at 700psi or 49kg/sq.cm. – the bridge could be opened and closed before the system needed re-pressurising, and four hours' notice was required if a ship needed to pass. Today it is 24 hours.

Remarkably, after 145 years of operation, much of Armstrong's hydraulic machinery is still used to rotate the bridge, although the accumulator sited to the east of the bridge is now charged by electricity rather than steam engines. The original design had two accumulators but the west accumulator is no longer used.

During the planning for their 1884 visit to Northumberland, the Prince of Wales announced that he and the Princess of Wales would be visiting Cragside. He might normally have been expected to stay at Alnwick Castle – home of the Duke of Northumberland – or at the home of another member of the nobility, but Cragside offered much more modern living conditions than could be found in Buckingham Palace – and he had apparently never experienced electric lights. Luckily he gave sufficient notice of his visit for Sir William to add a lavish new wing to the house especially for his royal guests. Amidst the celebrations of Queen Victoria's Golden Jubilee in 1887, Sir William became the first Baron Armstrong of Cragside. He died in 1900 aged 90.

At its peak, his company employed more than 25,000 people across several manufacturing sites, manufacturing hydraulic cranes and armaments, building ships and assembling steam and hydraulic engines. It had even briefly manufactured railway locomotives in its early days.

Long after Baron Armstrong's death, the Elswick Works was even contracted to build a consignment of 100 Mark IV tanks for the army, out of a total order of 1,220 spread across six different manufacturers.

[above left] Tower Bridge opens to accommodate the tall mast of the Thames Spritsail Barge *Gladys*, built in 1901 by Cann, John & Herbert of Harwich. She was recently restored at T. Nielsen & Co. in Gloucester.

[above] The steel-framed Leamington Lift Bridge on the Edinburgh and Glasgow Union Canal was built by W. G. Armstrong & Company in 1906 and originally sited at Fountainbridge closer to Edinburgh city centre and was moved to its present position in 1922. Out of use by the 1960s, it was fully restored in 2000–2001 as part of the £85M Millennium Project to restore the canal link between Edinburgh and Glasgow which had been lost with the building of the M8 motorway.

[top left] A replica of a Mark IV tank – built for the film *War Horse* – is displayed at the Tank Museum in Bovington, Dorset. Seven original Mark IVs survive, one of which is also displayed at Bovington.

[top right] The cramped gunnery position inside a World War 1 Mark IV tank. The contract for building 100 of them was given to W. G. Armstrong, Whitworth & Co. at Elswick in 1917.

[above] The Armstrong Whitworth maker's plate on a Hotchkiss six-pounder gun, manufactured under licence by Armstrong and fitted into one of the Mark IV tanks.

Armstrongs had been manufacturing Hotchkiss long-barrel six-pounder guns for the navy under licence for many years, and could quickly modify production to build the short-barrelled version which was fitted in the side sponsons of the tanks. The large-scale production facilities at the Elswick Ordnance Works were ideally placed to arm not only the tanks which they were building themselves, but perhaps they might even have supplied some of the guns for the tanks being built elsewhere.

Many others developed hydraulic systems – Edward Leader Williams' Anderton Boat Lift in Cheshire was originally raised and lowered by hydraulic rams when it first opened in 1875. Just 30 years later its hydraulics were abandoned and replaced by electric winches, but early this century it was restored to hydraulic power.

In William Jessop's Floating Harbour in Bristol, Armstrong hydraulics were installed to power the massive lock gates across Brunel's South Entrance Lock – previously manually operated – in order to give larger ships access to the docks. The hydraulic system was also used to power his 1849 Swivel Bridge which spanned the entrance. It had hitherto also been manually opened and closed.

The original hydraulic power system was completed in 1887 at the instigation of William Howard, the dock's first full-time engineer.

The steam-powered hydraulics were housed in William Howard's Cumberland Basin Hydraulic Engine House – now a pub and restaurant – but when steam was replaced by 365v electrical power in 1907, a new Pump House was built at Underfall Yard using pumping engines by Paisley-based engineers Fullerton, Hodgart & Barclay.

The pumps drew water from the harbour, pressurised it to 750psi, and used it to raise a giant accumulator – just as at Tower Bridge. That stored the energy to await demands on the system, and when power was required, the giant cylinder containing 80 tons of scrap metal was allowed to drop slowly under its own weight, forcing the pressurised water through the system to operate the dock gates and cranes.

Between 1901 and 1902 the system was connected to new rams in pits to operate Brunel's 1849-built Swivel Bridge over the Entrance Lock, again using Armstrong equipment. Before the first hydraulic system was installed in 1887, Brunel's gates were operated by dockhands. Today, hydraulics play their part in just about every machine from huge cranes to small robotic arms.

Armstrong, Mitchell & Co built their last warship in 1918 – the aircraft carrier HMS *Eagle* – but after the Great War, the Armstrong-Whitworth company went into a period of slow decline, and in 1927 the facilities at Elswick and Openshaw – the original Armstrong and Whitworth sites respectively – were sold, becoming known as Vickers Armstrong Limited. The former Mitchell yard continued to operate for a time as Armstrong, Mitchell & Co.

By 1935 Vickers Armstrong was the third largest manufacturing company in the country with shipyards on both the north-west coast – at Barrow-in-Furness – and the north-east coast. Armstrong Mitchell & Company went into voluntary liquidation in 1956 and, through a series of mergers and nationalisations during the 1960s and 1970s the Armstrong name eventually disappeared after more than a century at the heart of British industry. The Barrow yard continues in operation today as BAe Systems Submarines, builder of the Royal Navy's nuclear submarines.

[above left] William Howard's original 1887 Cumberland Basin Hydraulic House which housed Armstrong's hydraulic system. The accumulator which pressurised the water was contained in the tower.

[above] The Swing Bridge open to allow the passage of the Elswick-built HMS *Victoria*. The 11,200 ton battleship – the first to be fitted with triple-expansion steam engines – was launched in 1887 and entered service in 1890. Just three years later she was rammed off Tripoli by HMS *Camperdown* and sunk. Her engines were built by Humphrys, Tennant & Company of Deptford who supplied engines for many of Armstrong's vessels. By this stage in their history, Sir W. G. Armstrong, Whitworth & Company also had two newer yards on the north side of the river, east of the Swing Bridge, which could handle much larger vessels.

[left] When opened on 17 July 1876 Armstrong's hydraulic-powered Swing Bridge over the Tyne was the world's longest swing bridge. Behind it is Robert Stephenson's 1849-built High Level Bridge. This postcard was published around 1904.

High Level and Swing Bridges, Newcastle-on-Tyne

IRON MANUFACTURE.
BESSEMER'S PROCESS.

PLATE 2.

FIG. 3.—PLAN OF
TUYERE.

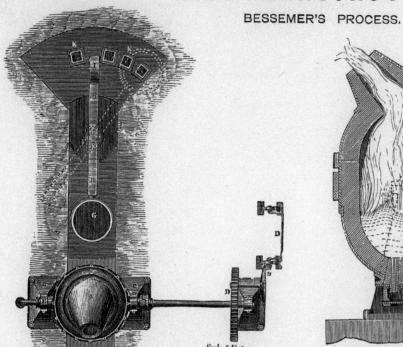

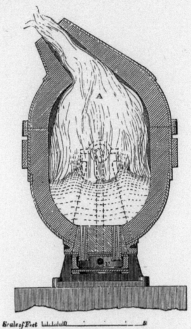

FIG 1.- PLAN.

FIG. 2.—SECTION OF CONVERTING
VESSEL.

FIG. 4.—VERTICAL AND
HORIZONTAL SEC-
TION OF TUYERE.

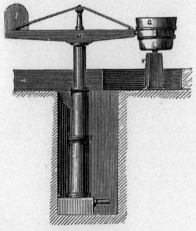

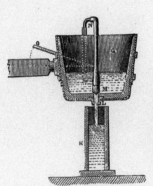

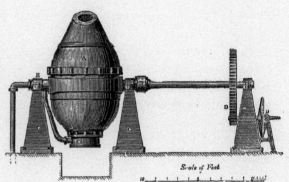

FIG. 5.—SIDE VIEW OF HYDRAULIC
CRANE RAISED.

FIG. 6.—SECTION OF LADLE
AND INGOT MOULDS.

FIG. 7.—FRONT ELEVATION.

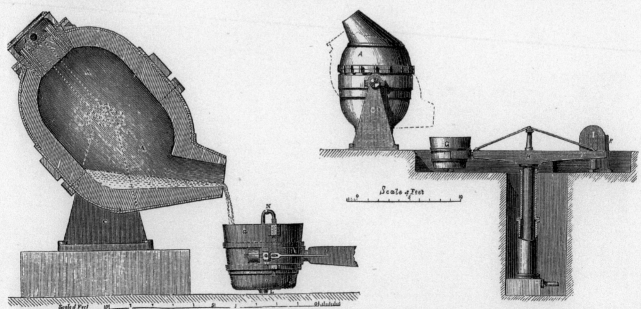

FIG. 8.—SECTION OF VESSEL, SIDE VIEW OF LADLE.

FIG. 9.—SIDE ELEVATION, PARTLY IN SECT.

SIR HENRY'S IRON CRUCIBLE

connections with:
- William Armstrong
- William Arrol

The Bessemer Converter revolutionised steel-making, not only improving the quality of the steel produced but also hugely reducing the cost of making it. The Converter was adopted to make 'Bessemer Steel' across the world and, with modifications and improvements, Bessemer Converters continued in use for more than a century.

Henry Bessemer (1813–1898) was born at the family estate near Charlton in Hertfordshire, purchased out of the proceeds of his father Anthony's inventions.

The family was descended from French Hugenots but his father, while having been born in London had moved to France becoming a celebrated inventor himself, eventually being made a member of the French Academie des Sciences at the age of just 26 in recognition of his work to improve the precision of optical microscopes. But all that counted for nothing when he was forced to flee the country at the time of the French Revolution.

Following in his father's footsteps, Henry also became a serial inventor – some of his inventions were successful, others less so, always driven by his belief that he could design and engineer a better way of doing something than anyone before him.

He made his first fortune by building a machine which could grind brass into very fine flakes – an essential ingredient in the manufacture of gold paint. The powder which was used to make gold paint at the time was, oddly, known in the trade as 'bronze powder' and was not manufactured anywhere in Britain. Thus in the 1840s picture framers and decorators were buying it from Germany at the substantial price of between £4 and £6 per pound. Bessemer's interest was triggered by buying just one ounce of it for a lettering job he wanted to complete for his sister, and being charged seven shillings, not far short of a week's wages for a labourer at the time.

His machine, once he overcame teething problems with it, apparently reduced the cost of manufacturing the powder to

[left] A caricature of Sir Henry Bessemer, from *Vanity Fair*, 6 November 1880.

[below] In its issue for 18 March 1916, the *Illustrated London News* published this photograph of Bessemer Converters at an un-named foundry, captioned 'An Early Phase in the Production of a Big Gun'.

[opposite page] Explaining the Bessemer Process for making steel visually – a plate from a late 19th century encyclopaedia.

[above] The Victorian gilt picture frame was ubiquitous, and most were coloured using gold paints made with the so-called 'bronze powder'. Bessemer reportedly sold an estimated 80,000 bottles of his 'gold paint' per year. This framed ambrotype portrait of a young child dates from the 1860s, by which time sales of Bessemer's paint were still returning him huge profits. In one of his obituaries, noting that it cost just a penny per pound to make the powder, the writer expressed some surprise that Bessemer had managed to keep the process secret for 40 years.

[below] This illustration of the SS *Bessemer* has disproportionately large and squat funnels nothing like the actual ship (*see the photograph opposite*) suggesting the artists from the *Illustrated London News* had never actually seen the vessel.

just 2.5% of what it had previously been, and in a canny bit of marketing, he offered to supply existing users of the powder at a substantial discount on what they were currently paying if they bought exclusively from him. His fortune was assured.

When he subjected it to chemical analysis, he found that 'bronze powder' actually contained no bronze whatsoever, instead being made entirely of tiny flakes of brass. He assumed that the price must reflect it being hand ground, and recognised that if he could build a steam-powered machine to make it, the rewards would be significant. A man of immense confidence, he once claimed that he had

'an immense advantage over many others ... ... inasmuch as I had no fixed ideas derived from long-established practice to control and bias my mind, and did not suffer from the general belief that whatever is, is right'.

He backed up that belief with more than 120 patents across a range of specialisms embracing military ordnance, an early typesetting machine, a means of compressing graphite to make better pencils, and several patents covering the sugar industry. And yet, he apparently had no formal education in engineering or the sciences, his experience in those subjects limited to working alongside his father.

A common theme throughout his working life seems to have been a mistrust of whoever he was dealing with, triggered by an early run-in with the British Stamp Office which produced the tax stamps used to validate contracts.

In 1833 Bessemer had suggested a cost-effective way of stopping people re-using stamps from earlier contracts – said to be losing the Government at least £100,000 pounds in revenue – by using a perforating machine to apply the 'stamp' as a pattern of 400 tiny holes directly on to the contract. Instead of negotiating royalties from his idea, he had accepted the post of 'Superintendent of Stamps' at the substantial salary of £700 per annum. His fiancée, however, suggested that the expense of such a machine was unnecessary, and that a much simpler approach was to date stamp the stamps themselves. Having offered her suggestion to the Stamp Office, without having first protected it, the authorities quickly realised they thus were free to use it without compensation – and without the need to employ Bessemer himself. The job and salary disappeared. He had learned a salutary lesson, aged just 20.

Thus, when he built his factory to make bronze powder, he did so in great secrecy, only trusting his three brothers-in-law with the details of the manufacturing process.

Amongst his many unusual inventions was a design for a ship which he claimed would not induce seasickness – Bessemer himself was prone to the condition and dreaded his journeys to and from France. So, in 1868, he drew up plans for his solution to the problem,

[left & below left] Two of the many published illustrations purporting to show how the self-levelling saloon on the SS *Bessemer* was intended to stay horizontal in a storm. The illustration, *left*, gives a sense of the vessel's lavish interior decoration. Of course the design could only counter a lateral roll.

[below] Such was the interest in the SS *Bessemer* – seen here newly arrived at Dover and about to embark on her maiden voyage across the Channel – that local Dover professional photographer Alexander James Grossmann produced 'cabinet prints' of the vessel and sold them from his studio in Snargate. He continued to operate a studio there until 1901.

and it certainly was original. The ship was to have twin engines, four paddle wheels, and a passenger cabin which used hydraulics and gimbals to keep it level as the vessel rolled from side to side.

It took several years before funding had been arranged through a number of backers, and a marine engineer had converted Bessemer's ideas into a viable design for what would turn out to be quite a large vessel.

It was 350ft (just under 107m) long with a beam of 65ft (just under 20m) across the paddle boxes. The Bessemer Saloon Ship Company Ltd ordered the 1,974grt vessel from the Earle Brothers shipyard in Hull, and it was launched in September 1874. A reporter from *The Times*, visiting the partly-built ship in Hull noted that:

[above] A Bessemer Converters at Leeds Steelworks in Hunslet, Yorkshire, c.1910. The original Leeds Iron & Steel Works had been taken over in 1900 by Walter Scott Ltd. At its peak, the works had four large blast furnaces making pig iron and at least two Bessemer Converters. In 1913 there was a major accident at the plant when a boiler exploded killing nine employees.

[above right] Carrying out essential maintenance on a Bessemer Converter in an American steelworks around 1900. Here the joint between bottom plate and shell is being repacked to seal the slight space between the two parts after the bottom has been latched back on. The photograph was taken by the New York based Underwood & Underwood studio. Lantern slides such as this were taken on black and white plates and individually and expertly hand-tinted – indoor colour photography was still some years in the future.

'Not only are the two ends of the ship alike, and furnished with rudders, but having for some thirty or forty feet from the extremities a low freeboard, they at present give to the vessel the appearance of a gigantic steam canoe. This aspect will, no doubt, disappear as progress is made with the central portion of the ship, and as her paddle boxes are erected; but in her finished condition the vessel will present an unexampled appearance at the extremities.'

Before work had started on building it, the concept had been tested using models, but sadly what worked well with models did less well at full size. On its first test crossing of the Channel in April 1875 – with just its crew on board to test the systems – it was found to be difficult to steer at slow speed, and damaged one of its paddle-boxes as it scraped the jetty on arrival at Calais.

A month later, with the paddle-box repaired and the self-levelling mechanism locked out of service, the ship made its first and only crossing with fare-paying passengers – this time crashing into and demolishing part of the Calais jetty. The self-levelling mechanism was never tried out, the backers withdrew their funding support and the vessel was consigned to the breakers' yard in 1879 after three years tied up at Dover. The flamboyantly furnished saloon, however was saved and given a second life – as a lecture room at the Kent Horticultural Institute in Swanley – until destroyed in a bombing raid in 1944.

Of all Bessemer's ideas and inventions, none equalled the impact on the industrial world which would be triggered by the introduction of his steel-making process. It is rare to describe an invention as 'world-changing', but Bessemer Steel was just that. At a stroke, good quality steel went from being a very expensive commodity to being affordable.

Bessemer did not invent steel, but like so many of his other inventions, he found a way of making it better, cheaper and in greater quantities. With his patent process, steel could be produced in truly industrial quantities for the first time, at a fraction of the cost and in a fraction of the time – a good example of his claim that he 'did not suffer from the general belief that whatever is, is right'.

The early processes for turning pig iron into steel can be traced back before mediaeval times, but they were slow, inefficient and not always successful.

Pig iron contained a lot of carbon and other elements and impurities and, in order to convert it into steel, the unwanted chemicals and minerals – which included manganese, silicon, sulphur, phosphorus and others – had to be modified or removed. The easiest way to do that was by what was referred to as the 'crucible method' where a cocktail of other chemicals was added to drive out the excess carbon and other unwanted elements from the molten metal.

The ratio of those constituents left after the melt played a major role in determining the character of the steel – high carbon steel was very hard but quite brittle; low carbon steel was more durable. Iron ores from different parts of the same country had different chemical make-ups, producing steel of a very different character than others. Cumbrian haematite, for example, was much higher in phosphorus than the ores mined around Blaenavon in South Wales.

While steel-making on a small scale had been known for centuries, the start of the Industrial Revolution quickly raised demand beyond capacity, and in the 18th century Benjamin Huntsman, a clockmaker, set out to produce a better steel for making clock springs, and was moderately successful. He began producing steel in 1740 after lengthy experimentation, using a coke furnace capable of reaching

a temperature of around 1,600°C. The furnace was loaded with up to 12 clay crucibles, each holding about 34lbs (15kg) of iron. After several hours sustained heat, a typical production total would have been around 400lbs (180kg) of a hard steel.

The small-scale production, the costly additives, the additional fuel involved and the labour-intensity of the process, meant that steel was a highly priced commodity. A survival from this period – using steel made by the Huntsman method – can be seen in the scythe-making works at Sheffield's Abbeydale Industrial Hamlet.

Bessemer's process involved filling his large egg-shaped 'convertor', which was lined with fire bricks, with molten iron, and then blasting air through it from the base. That set off a spectacular reaction and it must initially have seemed almost counter-intuitive – instead of that air cooling the molten iron, it triggered an exothermic reaction, increasing the temperature of the melt considerably, burning off more of the carbon and creating a purer mild steel.

On 11 August, 1856, he outlined his process in a paper, titled *The Manufacture of Malleable Iron without Fuel*, to the British Association at Cheltenham and immediately started selling licences. Unfortunately, the process did not work easily with some types of pig iron, and getting the appropriate flow rate and pressure of the air being passed through the molten iron proved difficult. As a consequence, Bessemer had to repay most of his investors.

After several years of further experimentation, during which time he leant heavily on the innovation of others, especially in Sweden, the glitches in the process were eliminated and he produced his first successful 'converter' in 1865. That is displayed in the Science Museum in London.

The secret lay in having more nozzles – known as 'tuyerès' – in the base of the converter, and in the steady delivery of a air at a precisely-controlled pressure. Once that was achieved, success followed.

The first converter was installed in the premises of the Workington Haematite Company – in which Bessemer had a significant interest – and by 1872, steel was being produced there

[top] The Bessemer Converter displayed at the Kelham Island Museum in Sheffield produced the last Bessemer steel in Britain at British Steel's Workington plant in Cumbria in 1974 and is one of only four survivors of the many built. The other examples are in Pittsburgh PA and Fagersta, Sweden. The 1855 prototype is in the Science Museum in London.

[above] Once the iron had been converted into molten steel it was drawn off into this large ladle – now also at Kelham Island – ready for casting into ingots.

[right] A view of the base plate of one of the converters at the Cambria Steel Plant in Pennsylvania shows the 24 nozzles through which the compressed air used during the 'blow' was introduced. When the converter's air jets – known as 'tuyerès' – in the bottom of the device needed cleaning or replacing, the whole of the lower assembly could be un-latched from the shell of the crucible. In this photograph from the early 20th century, the entire set of tuyerès is being replaced. (courtesy of the Hagley Museum and Library, Wilmington, Delaware)

[below] An item of paperwork from the winding up of the Bessemer Steel and Ordnance Company in 1874.

in a trio of converters with a total capacity of 24tons. The ability to produce such a quantity of steel in just 30 minutes was a 'game-changer', but within a few years converters each with a capacity of 25tons each were commonplace.

It is an interesting anecdote to the Bessemer process that the last working converter in Britain was withdrawn in 1974 from that same Workington steelworks, by then operating as part of the British Steel Corporation.

One of the many companies Bessemer established to operate his converters was the Bessemer Steel & Ordnance Company of Gracechurch Street in London. The company, formed in 1871 with some influential shareholders, built a new steelworks in East Greenwich, but by 1874 it was in administration – despite growing demand for Bessemer Steel it was just too far away from the raw materials it needed.

Having not achieved the immediate returns he had expected as a steel producer, a licensing model proved much more lucrative, and within a few years, Bessemer Converters – with several times the capacity of the original – were in use all over the world.

Other engineers – especially in Sweden and the United States – modified and improved the device over the years, increasing the purity of the steel it could produce, reducing the 'blow' time, and developing methods for producing different types of steel. That was achieved, perversely by re-introducing some of the elements which had been burned off in the 'blow'.

In May 1891, the French engineer Alexandre Tropenas filed a provisional patent specification for an evolution of the Bessemer Converter, and his patent was granted in April 1892 under the title '*Improvements in the Manufacture of Steel, Steel Castings, or Ingot Iron, and in Apparatus and Appliances employed therein.*'

Whereas the Bessemer process involved blowing air through the molten metal in the converter to burn off impurities, the Tropenas process blew super-heated air through two banks of tuyères across the top of the molten iron. With a pressure of just 5psi, the effect of the hot air blast across the surface of the melt was sufficient to increase its temperature and thus burn off even more of the impurities in it.

Our world owes its existence to the availability of the vast range of high quality steels which are manufactured across the world today. Each different steel has its own particular characteristics designed for a specific use. The manufacture of modern steels, while using the same raw materials that Bessemer used, has progressed a long way from his original process, the final characteristics of each material being defined by engineering precise chemical compositions.

In an obituary published by the Iron and Steel Institute at the time of Bessemer's death, the importance of his contribution to 19th century industrial advancement was ranked alongside that of James Watt.

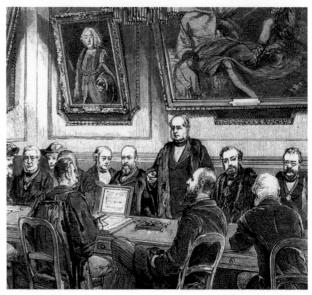

'Sir Henry Bessemer died, at the age of eighty-five, at his residence at Denmark Hill, on March 15, 1898. By his death British metallurgy has to deplore the loss of one whose name will be for ever associated with the record of its progress and development, as that of Watt with the steam-engine. The discovery of the means of rapidly and cheaply converting pig iron into steel by blowing a blast of air through the molten iron was the result of labours and experiments that extended over a period of more than ten years, the results being attained only after many disheartening failures. Prior to this invention, the entire production of steel in Great Britain did not exceed 50,000 tons annually, and its price, which ranged from £50 to £60 per ton, precluded its use for most of the purposes to which it is now universally applied. At the present time the world's production of steel made by the Bessemer process has reached the enormous total of 11,215,000 tons per annum, Great Britain alone producing 1,845,000 tons. Its selling price has been reduced to about £4, 10s. per ton. No other invention has had such remarkable results. It is no exaggeration to say that without the Bessemer process for steel rail-making the present railway system of the world would not now exist.'

[above] The last surviving Tropenas Converter in Britain is displayed at the Museum of Scottish Industrial Life in Coatbridge.

[below] In recognition of his many achievements, in 1880 Bessemer was presented with the Honorary Freedom of the Worshipful Company of Turners – one of the City of London's oldest Livery Companies dating back to 1604.

Obituaries and eulogies of great men and women invariably overstate their contribution to the world, but in Bessemer's case, no exaggeration was necessary. His legacy still impacts on our lives today.

WILLIAM ARROLL, MAN OF STEEL

Watching one of Britain's most iconic vessels pass beneath one of Britain's most iconic bridges is a memorable experience. The Clyde-built PS *Waverley* and Sir William Arrol's Tower Bridge are both powerful evidence of the quality and enduring status of British engineering prowess.

The design of Tower Bridge was arrived at as a result of the challenges faced by Victorian engineers of bridging the Thames at that point – there had to be clearance for tall-masted ships at all stages of the tide, for the heart of Victorian London's dockland was then on the upper river. No fewer than 50 possible designs were proposed, including one for the 'Bascule Bridge' we see today, proposed by Sir Horace Jones, modified and engineered by John Wolfe Barry – the son of Sir Charles Barry who had designed and built the Houses of Parliament – and Henry Marc Brunel, Isambard Kingdom Brunel's younger son.

The steelwork was contracted to William Arrol (1839–1913) – he became Sir William during the project – one of the many to fully exploit the availability of low-cost Bessemer steel. By the time he was awarded the Tower bridge contract, William Arrol, a spinner's son born in the village of Houston in Renfrewshire, had risen to be one of the greatest engineers of Victorian and Edwardian Britain, establishing a company which would build some of the country's most immediately recognisable structures. Interestingly, Tower Bridge is one Arrol project where very little of the company's input is actually evident from the outside – but take the tour inside the towers, and behind the beautiful stone cladding is a complex steel structure more than able to withstand the stresses the bridge has endured in its century and a quarter spanning the Thames.

The bridge is a monumental structure, and the statistics of its construction are impressive. Originally budgeted at around £750,000, the final bill was just short of £1,200,000, so cost

[above] Sir William Arrol.

[below] Inside Tower Bridge, Arrol's latticed steelwork can be appreciated.

[opposite page] The paddle steamer *Waverley* – Britain's last sea-going paddler – passing underneath London's iconic Tower Bridge.

connections with:

- William Armstrong
- Henry Bessemer

overruns on major infrastructure projects are not just a modern phenomenon. Its features were technically innovative, posing considerable engineering challenges to its builders, not least being the huge weights involved.

Construction started in 1887, took seven years, employed over 400 workmen, and drew crowds of visitors as work progressed. The bridge was finally opened by the Prince and Princess of Wales on 30 June 1894, allowing vessels of up to 10,000 tons to pass between the towers.

Six months before the official opening, the journal *The Engineer* considered the bridge to be a project of such interest that it devoted 33 pages of its 15 December 1893 issue to a detailed description of the bridge and its innovative construction.

Beneath the granite-cladding of the towers which we see today is a complex steel-framed structure standing on huge concrete piers each weighing 70,000 tons. That weight was required to support the bridge itself and deal with the stresses as the counter-balanced bascules raised and lowered the roadway. At the time the piers were built, they were the largest and heaviest in the world.

In total, 37,000 tons of concrete, 27,00 tons of bricks, 30,000 tons of masonry and 20,000 tons of cement were used in its construction, along with 13,000 tons of steel, wrought iron and cast iron. The wrought iron was used for the caissons which protected the workers as they excavated the river bed before constructing the piers.

The steelwork for the internal frame was manufactured in Glasgow – some of it pre-assembled at Dalmarnock into five-ton sections – and shipped almost half way round Britain, before being transferred to barges for the last half mile to the site.

On-site assembly was accelerated by the use of Arrol's patented steam-powered riveter – a device he had developed during the construction of the Forth Bridge. Mechanically, at the heart of the bridge were two massive horizontal twin-tandem compound stationary steam engines, built at Sir William Armstrong, Mitchell & Company's Elswick Works in Newcastle.

William Arrol was born in February 1839, two years after Queen Victoria came to the throne and, from leaving school at the age of nine, he worked in the same cotton mill as his father before being apprenticed to a blacksmith in Paisley.

He then took on a number of posts in engineering workshops in the area, and by 1863 he had joined Glasgow engineers and bridge builders R. Laidlaw & Son – who advertised themselves as 'Engineers & Contractors, Iron & Brass Founders' – at their Barrowfield Ironworks in Lambhill. Laidlaws were, at the time, described as 'one of the largest establishments of the kind in Glasgow' with a heritage already stretching back nearly 80 years, so it was an ideal environment in which to learn his trade.

Five years later, in 1868, he set up on his own as a boiler-maker in Bridgeton – using his entire savings of £85 to fund the new venture – before founding his eponymous company in 1871 at the age of 32 – William Arrol & Company – with their Dalmarnock Iron Works at Dunn Street in Dalmarnock, Glasgow. The new company's first major projects included the construction of the bridge across the Clyde at Bothwell for the Caledonian Railway Company and, for the same company, another railway bridge, completed in 1878, across the river at the Broomielaw in Glasgow.

That latter bridge was dismantled in the 1960s. The construction of a second bridge alongside it, built between 1899 and 1905 was also contracted to Arrols. For a young company to have been entrusted with such important projects suggests William Arrol had rapidly built up a significant reputation. Within a few years, Arrols had been appointed as the main engineering contractor for a proposed four-tower suspension bridge across the Firth of Forth which Thomas Bouch had designed to carry railway lines. Bouch's design, which had two towers on the island of Inchagarvie, shared a lot of its ideas with one of James Anderson's 1818

Dec. 15, 1893.

THE ENGINEER.

563

THE MIDDLESEX MAIN PIER AND HIGH LEVEL FOOTWAYS—VIEW FROM SOUTH (From Photographs by Mr. W. E. Wright) THE SURREY MAIN PIER FROM THE SOUTH

proposals for a road bridge across the river. He had suggested two designs – one a suspension bridge, the other a cable-stayed bridge not unlike the Queensferry Crossing a few hundred yards upstream which opened in 2018.

Bouch's contract was quickly cancelled after the Tay Bridge disaster in 1879, and a new and much stronger cantilevered design by Benjamin Baker and John Fowler was developed. While Bouch's reputation had been ruined by the Tay Bridge collapse, his Forth Bridge design would almost certainly have fared better, for it was to be built of steel rather than the cast iron and wrought iron which had been the case with the failed Tay Bridge. Arrol, already one of the leading proponents of steel construction, had been Bouch's first choice as engineer for the Forth crossing, and he was also Baker and Fowler's first choice.

By the time work began on the Forth Bridge, Arrol was also heavily involved in the building the replacement Tay Bridge which had been designed by early steel pioneer William Henry Barlow. That would prove to be much stronger than the original, and with double rather than a single track. It still carries the main East Coast railway line today more than 130 years after it was opened to traffic. Contrary to what many people think, not all the ironwork of the original bridge was of poor quality, and much of it was actually incorporated into the new bridge. The new bridge was opened in 1887, just three years before construction work was completed on the Forth Bridge.

During the years when construction of the two bridges overlapped, Arrol's workload was phenomenal. He would rise at 5am, arriving at Dalmarnock by 6am and spending a couple of hours checking progress and quality before taking the train to Edinburgh and spending Monday and Tuesday on site supervising work on the Forth Bridge. On Tuesday evenings he would take another train – and ferry – to Dundee. Wednesday was spent in the Tay Bridge site

Two of William Edward Wright's many progress photographs published in *The Engineer*'s thirty-three-page feature on the design and construction of the bridge.

[above] The Forth Bridge's construction was photographed regularly by several photographers – including James Valentine of Dundee and George Washington Wilson of Aberdeen. The official photographer, however, was Evelyn Carey, whose images form a unique and outstanding record of one of the late 19th century's major engineering projects.

[right] The Forth Railway Bridge – photographed from North Queensferry. In front of the central (Inchgarvie) tower, a small beacon stands on the base of one of the piers for Thomas Bouch's planned bridge, work on which had started in 1874 only to be abandoned five years later when his Tay Bridge collapsed.

office before a late night return to Glasgow, spending the rest of the week at Dalmarnock overseeing the construction of the prefabricated sections of steelwork for both bridges.

Arrol was an innovator, and had a reputation for being able, very quickly, to design a piece of equipment specifically to resolve any problems found during the Forth Bridge's construction. The pneumatic shovels used to speed up the process of digging compacted mud and other materials out of the caissons as the bases for the bridge's piers were constructed is a case in point. If a problem needed solving, Arrol could usually solve it.

One of his most innovative designs was for a portable hydraulic riveter which greatly speeded up the repetitive task of fixing the rivets – an estimated total of more than seven million – which hold the structure together.

After the Tay Bridge debacle, safety concerns were paramount in the design and construction of the Forth Bridge. As a result it was heavily over-engineered, but has stood the test of time – a structure well worthy of its World Heritage Site status. After the safety inspections had been carried out prior to the bridge's opening, the Board of Trade report included a comment which could have been said about every project Arrol undertook, noting that

> 'this great undertaking, every part of which we have seen at different stages of its construction, is a wonderful example of thoroughly good workmanship with excellent materials, and both in its conception and execution is a credit to all who have been connected with it.'

Within just seven years, Arrols had seen the completion of three massive bridges – the replacement Tay Bridge opening in 1887, the Forth Bridge in 1890, and Tower Bridge in 1894. And if that had been all the great man achieved, he would have more than earned his reputation, but there was so much more.

By the banks of the Clyde, stands another iconic Arrol structure – the giant cantilevered crane which is all that remains of the world-famous John Brown shipyard where some of the world's most famous ships were built – the Cunarders *Queen Mary*, *Queen Elizabeth* and the beautiful *QEII* to name but three.

In the company's early years, Arrols had acquired the Parkhead Crane Works in Glasgow's Rigby Street, and there they developed some of the biggest shipyard and dockside cranes in the world. In Glasgow, just two of their giant cantilevered cranes survive, and the largest is now open to the public.

The huge Arrol 'Titan' crane, 150ft high, was completed in 1907 and stands alone in the otherwise demolished shipyard. It was the world's first Giant Cantilever Crane. The crane was originally built to lift 150 tons, was modified in 1938 to carry loads of up to 200 tons. That modification involved the insertion of a considerable amount of extra steel-work to strengthen the tower and it became the biggest crane on the Clyde.

Three other big cranes survive on the river – the 1931-built Stobcross, or Finnieston, Crane had a lifting capacity of 175 tons, while the James Watt Dock crane in Greenock – built in 1917 – and 1920-built Barclay, Curle & Company crane at Whiteinch could both lift 150 tons. Both the Greenock and Whiteinch cranes were also built by Arrol. While the foundations of the Finnieston crane were constructed by Arrol, the tower was erected by Cowans, Sheldon & Company of Carlisle and the cantilever constructed by the Cleveland Bridge & Engineering Company. Another Arrol crane – at the Buccleugh Dock in Barrow-in-Furness – was completed in 1908 or 1909. It was destroyed during the Blitz and replaced with one of similar design, also from Arrols. In 2010 that too was demolished by BAe Systems.

The Clydebank crane cost £24,600 to build when completed in 1907, but in excess of £3,750,000 to restore as a visitor attraction a century later. Sadly, the surrounding landscape is derelict, so the crane is not displayed as well as it ought to be.

Relatively few of these giant cranes were ever built, with Arrols building 14 and being involved in the design of several others – including the monster at the Garden Island Naval Base in Sydney, Australia, which was demolished in 2015.

The example which was built in 1909 by Arrol for the North Eastern Marine Engineering Company in Wallsend, with machinery by leading crane engineers Stothert & Pitt of Lower Bristol Road in Bath, briefly enjoyed Grade 2* listed protection, having been identified as

[below left] Wormit Station and the Tay Bridge, from an Edwardian postcard, showing the bases of the piers from Bouch's bridge.

[below] One of the series of photographs by Valentines of Dundee taken in the 1880s which were later re-published as Edwardian postcards. To the right is part of one of the surviving piers from Thomas Bouch's ill-fated bridge while, to the left, the new bridge is nearing completion.

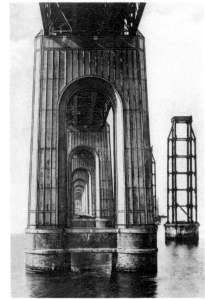

[clockwise from top left]
The giant cantilevered
crane at the former
John Brown shipyard,
Clydebank. It is known
locally as 'the Big Cran'.

The 1909 Arrol crane in
the North Eastern Marine
Engineering Company's
Wallsend yard was said to be
the largest in the world – it
stood perhaps two feet taller
than the Clydebank Titan.

The Garden Island Naval
Base crane in Sydney,
Australia, was designed
by Arrol but demolished
in 2015.

being of significant national importance in 1989. That protection, however, was withdrawn the following year and the crane demolished shortly afterwards. Stothert & Pitt had, themselves, been manufacturing cranes since 1869.

After years of debate, Middlesbrough Council in the north-east of England gave Arrols the contract to build their massive cantilevered transporter bridge over the River Tees linking Middlesbrough with Port Charlotte. Like many other Arrol structures it is still doing what it was designed to do more than a century ago. A stone plaque at the Middlesbrough end of the bridge records the work of Sir William Arrol & Company – Arrol had been knighted by Queen Victoria in 1890 for his services to engineering.

Building a transporter bridge across the Tees at Middlesbrough had first been proposed by Charles Smith in 1873 and 38 years after Smith published his design, just such a bridge was opened with great ceremony by Prince Arthur of Connaught. It even looked a lot like Smith's bridge.

The bridge was actually designed by the Cleveland Bridge & Engineering Company, who were disappointed not to get the construction contract, but Arrol had managed to undercut all the other bidders. It remains a moot point whether or not the Arrol quote was ever realistic. The final bill, at £87,300, was £19,000 – a massive 28% over budget.

The steel fabrication was undertaken at Arrol's Dalmarnock Works in Glasgow and transported down to the site – surprising given the ready availability of high quality steel on Teeside. More than 2,800 tons of steel went into building the bridge and it has been roughly estimated that somewhere around 4,000,000 rivets were used in the process – almost two-thirds of the 6,500,000 said to have been used on the 8,094ft long (2,529 metres) Forth Bridge.

[left] The Tees Transporter Bridge, photographed in the 1960s from a vessel on the river.

[below] The official opening of the Tees bridge in 1911.

[bottom] The second Transporter bridge which linked Joseph Crosfield's two industrial sites on either side of the River Mersey in Warrington, Cheshire

Construction started in July 1909 and took 27 months. There was an incentive for Arrols to complete the project on time as the contract stipulated a penalty of £50 per week for any overruns. The actual building programme was subject to several modifications introduced to increase the bridge's carrying capacity which, in turn, required a rethink on the strength of the main beam.

That need not have been an issue as, like many Arrol projects, the transporter was considerably over-engineered. More than a century later, its complex cantilevered beam remains relatively true – with just a very slight kink evident towards the Port Clarence side and it experiences only a tiny deflection as the fully-laden gondola reaches the mid point of its traverse. More than a third of that project build time, and not far short of 40% of the cost, was taken up with the excavation of the foundations for the tower piers and the building of the concrete base piers on either side of the river.

Arrol's other transporter bridge contract was for the second of two transporters to cross the Mersey at Warrington, linking Joseph Crosfield & Sons' two soap and chemical works on opposite banks of the river. The first bridge, a lightweight structure built by Thomas Piggott & Company, was in use by 1905 but as the chemical works expanded, work was started on a second and much stronger bridge in 1911.

PRINCE ARTHUR OPENS "TRANSPORTER BRIDGE" 17.10.1911.

With a span of just 187ft (57m), the second bridge was completed in 1915 and still stands today. It is the only transporter in the world specifically built to carry railway wagons, although later modified to carry road vehicles as well. Also known as Crosfield's Bridge or Bank Quay Bridge, it was last used around 1964 and is now classed as being one of many historic buildings on the 'at risk' register. The 'Friends of Warrington Transporter Bridge' are working hard to raise public awareness of its condition before it is too late to save it.

LAMBERT & BUTLER'S CIGARETTES

15 H.P. NEW ARROL-JOHNSTON CAR.

By 1896, the forward-thinking Sir William broadened his interests significantly when he joined forces with George Johnston to establish the Arrol-Johnston car company in Paisley, and effectively start Scotland's motor industry – although the origins of Scottish-built road vehicles can actually be traced back to William Murdoch's demonstration of a steam-powered carriage in 1784, but it was the production of the first Arrol-Johnston which really established a motor industry. In demonstrating the power of the car, Johnston achieved a speed of 17 miles per hour, and was promptly fined 'half a crown' – 2/6d, 12.5p – for breaking the law. Fourteen miles per hour was the maximum allowed under the newly-introduced 1896 *Locomotives on Highways Act*.

Arrol-Johnston cars were held in such high regard that when Ernest Shackleton was planning his 1908 'Nimrod' expedition to the South Pole, he planned to take a motor car with him in addition to the normal complement of horses, sledges and dogs.

The chosen car was based on the company's 1907 15hp open tourer, heavily modified to cope with the extreme conditions the explorers were likely to encounter. Some of the modifications would later find their way into production cars. The air-cooled engine was fitted with what were described as 'anti-freezing arrangements' while passenger comfort was enhanced by diverting exhaust gases into pipework in the footwell to serve as foot-warmers. The gases were also used to melt snow in a tank on one side of the car, creating a supply of fresh drinking water. The rear wheels were replaced by stronger ones fitted with heavy-duty tyres fitted with iron spikes on them, while the front wheels were clamped on to skis. The company later moved to Dumfries and continued to produce cars into the early 1930s. Amongst their Dumfries models, the 'Galloway' was described as a car built by women for women to drive.

While many magnificent monuments to the Arrol company's achievements still survive, their greatest 'lost' achievement was the huge 'Arrol Gantry' which was designed and was built for the Belfast shipyard of Harland & Wolff. This structure was so large that the great White Star liners *Titanic, Olympic* and *Britannic* were constructed almost completely within its berths. Built from 6,000 tons of steel, it was 840ft (256m) long, 260ft (82m) wide and nearly 230ft (70m) high. It was completed in 1908, with work starting on *Olympic* almost immediately. It remained in use into the 1960s when it was dismantled during a major reorganisation of the yard.

Arrol's company continued in business long after its founder's death, and was a member of the consortium which built the Forth Road Bridge in the 1960s. His engineering legacy can still be seen all over the world.

[this page above] The huge Arrol Gantry at Harland & Wolff's Queen's Island Shipyard was used during the construction of three White Star liners. The smaller Berth 1 is to the left of the picture.

[left] The RMS *Britannic* (Hull No.433) under construction in Berth 2 in 1914.

[bottom] RMS *Olympic* (Hull No.400) is in Berth 2 beneath the right-hand gantry with RMS *Titanic* (Hull No.401) in Berth 3 beneath the left-hand one, from a chromo-lithographed postcard published in 1909. In reality, RMS *Titanic* was painted black during construction while *Olympic* was painted white for the benefit of photographers and repainted in White Star livery just before launch. Berth 1 can be seen in the distance to the right of this picture.

The World's Greatest Gantry, in Harland and Wolff's North Shipyard, Belfast.

[opppsite page top] A Paisley-built Arrol-Johnston 15.9hp car introduced in 1911.

[bottom left] A 1927 1.5 litre Arrol-Johnston saloon.

A cigarette card of Shackleton's Arrol-Johnston car which went on the South Polar Expedition. It proved too heavy to work efficiently in snow, but handled well on sheet ice.

The maker's plate on a preserved 'Galloway' car.

FURTHER READING

Bailey, Michael R., *Built in Britain – The Independent Locomotive Manufacturing Industry in the Nineteenth Century,* Canal and Railway Historical Society, 2021

Baker, Alan C. & Fell, Mike G., *Harecastle's Canal and Railway Tunnels,* Lightmoor Press, 2019

Bathurst, Bella, *The Lighthouse Stevensons,* Harper Collins, 1999

Binding, John, *Brunel's Royal Albert Bridge,* Twelveheads Press, 1997

Beckett, Derrick, *Telford's Britain,* Simpkin, Marshall & Co, 1859

Broom, Ian, *The Crofton Story,* Wiltshire Archaeological and Natural History Society, 2013

Brunel, Isambard, *The Life of Isambard Kingdom Brunel,* David & Charles, 1987

Burrow, J. C, and Thomas, William, *'Mongst Mines and Miners – Underground Scenes by Flash-Light,* Simpkin, Marshall, Hamilton, Kent & Co, 1893

Byrom, Richard, *William Fairbairn, The Experimental Engineer,* Railway and Canal Historical Society 2017

Clark, D. Kinnear, *The Steam Engine – a Treatise on Steam Engines and Boilers,* Blackie & Son, 1890

Corbie, Nick, *James Brindley – the first canal builder,* Tempus, 2005

Corlett, Ewan, *The Iron Ship,* Brunel's SS Great Britain Trust, 2012

Cross-Rudkin, Peter, *John Rennie 'Engineer of many splendid and useful works',* Railway & Canal Historical Society, 2022

Dawson, Anthony, *The Liverpool and Manchester Railway: An Operating History,* Pen and Sword, 2020

Demidowicz, George, *The Soho Manufactory, Mint and Foundry, West Midlands,* Historic England/Liverpool University Press, 2022

Dick, Malcolm and Croft, Kate, *The Power to Change the World, James Watt – a life in 50 objects,* West Midlands History Ltd, 2019

Emmerson, George S., *John Scott Russell – a Great Victorian Engineer and Naval Architect,* John Murray, 1977

Evans, Kathleen, *James Brindley – Canal Engineer,* Churnet Valley Books, 1998

French, Gilbert J., *Life and Times of Samuel Crompton,* Simpkin, Marshall & Co, 1859

Fairbairn, William, *On tubular wrought iron cranes: with description of the 60 ton tubular wrought iron crane recently erected at Kayham dockyard, Devonport,* The Institution of Civil Engineers, 1857

Fairbairn, William and Pole, William, *The Life of Sir William Fairbairn, Bart,* Longmans, Green & Co, 1877

A Bessemer Convertor in action, painted by Charles de Lacy (1856–1929) as an illustration for the 1930 *Waverley Book of Knowledge*. The caption reads 'Our Age is built on iron, almost always in the form of steel. A Bessemer Converter will make 20 tons of steel in a single 'blow' of 15 or 20 minutes. A huge iron cauldron, lined with fireclay and bricks, bows its head to be fed with its white-hot ration of molten iron, then rears up on its haunches, growling and sending forth a great blast of dazzling flame as air is forced through from the bottom at high pressure. The result is steel, the giant of modern industry, which is run off in a molten state. The Converter to the left has finished its work, while its companion is just beginning.'

Fairbairn, William *Treatise on Mills and Millwork, (Fourth Edition)*, Longmans, Green & Co, 1878

Gibson, Mike, *Pennine Pioneer: The Story of the Rochdale Canal*, History Press, 2004

Glover, Julian, *Man of Iron – Thomas Telford and the Building of Britain*, Bloomsbury, 2017

Griffiths, John, *The Third Man – the life and times of William Murdoch, inventor of gaslight*, André Deutsch, 1992

Hadfield, Chas & Skempton, A. W., *William Jessop, Engineer*, David & Charles, 1979

Hannavy, John, *Transporter Bridges – an illustrated history*, Pen and Sword, 2020

Hannavy, John, *The Governor – controlling the power of steam machines*, Pen and Sword, 2021

Hills, Richard, *Power from Steam*, Cambridge University Press, 1989

Kelly, Maurice, *The Non-Rotative Beam Engine*, Camden Miniature Steam Services, 2002

Malpass, Peter & King, Andy, *Bristol's Floating Harbour – the First 200 years*, Redcliffe Press, 2009

Marsden, Ben, *Watt's Perfect Engine*, Icon Books, 2002

Matthews, William, *An Historical Sketch of the Origin and Progress of Gas-Lighting*, Simpkin & Marshall, 1832

Meyer, Joseph, *Konversations-Lexikon*, Bibliographisches Institut, 1885

Morrison-Low, Alison, *Northern Lights – the age of Scottish lighthouses*, NMSE Publishing, 2010

Nancollas, Tom, *Seashaken Houses*, Particular Books, 2018

Rees, Abraham, *The Cyclopædia, or, Universal Dictionary of Arts, Sciences, and Literature*, Longman, Hurst, Rees, Orme & Brown, 1820

Reyburn, Wallace, *Bridge Across the Atlantic*, Harrap, 1972

Smiles, Samuel, *The Life of George Stephenson*, John Murray, 1857

Smiles, Samuel, *Lives of the Engineers*, John Murray, 1862

Smiles, Samuel, *Industrial Biography – Iron Works and Tool Makers*, John Murray, 1863

Smiles, Samuel, *Lives of Boulton and Watt*, John Murray, 1865

Smiles, Samuel, *Men of Invention and Industry*, John Murray, 1884

Smiles, Samuel, Editor, *James Nasmyth, An Autobiography* John Murray, 1885

Southey, Robert, *Journal of a Tour in Scotland in 1819*, John Murray, 1929

Tangye, Richard, *"One and All", an Autobiography of Richard Tangye*, S. W. Partridge & Company 1902

Wakelin, Peter, *Pontecysyllte Aqueduct and Canal*, Canal and River Trust, 2015

Waller, David, *Iron Men, how one London factory powered the Industrial Revolution*, Anthem Press, 2016

Winchester, Simon, *Exactly – how precision engineers created the modern world*, William Collins, 2018

Young, Charles F. T., *The Fouling and Corrosion of Iron Ships: Their Causes and Means of Prevention*, The London Drawing Association, 1867

INDEX